中英双语

大爆炸小史

A Little Book About the Big Bang

［英］托尼·罗思曼　著

陈冬妮　译

中国科学技术出版社

·北　京·

图书在版编目（CIP）数据

大爆炸小史 /（英）托尼 · 罗思曼著；陈冬妮译 . -- 北京：中国科学技术出版社，2024.3

书名原文：A LITTLE BOOK ABOUT THE BIG BANG

ISBN 978-7-5236-0320-8

Ⅰ. ①大… Ⅱ. ①托… ②陈… Ⅲ. ①宇宙—普及读物 Ⅳ. ① P159-49

中国国家版本馆 CIP 数据核字（2023）第 218886 号

策划编辑	高立波　夏凤金
责任编辑	夏凤金
封面设计	北京潜龙
内文设计	中文天地
责任校对	邓雪梅
责任印制	李晓霖

出　　版	中国科学技术出版社
发　　行	中国科学技术出版社有限公司发行部
地　　址	北京市海淀区中关村南大街16号
邮　　编	100081
发行电话	010-62173865
传　　真	010-62173081
网　　址	http://www.cspbooks.com.cn

开　　本	880mm × 1230mm　1/32
字　　数	364千字
印　　张	11.25
版　　次	2024年3月第1版
印　　次	2024年3月第1次印刷
印　　刷	北京长宁印刷有限公司
书　　号	ISBN 978-7-5236-0320-8 / P · 226
定　　价	78.00元

献给我的老师和同事们，

他们教给我的远比他们知道的更多。

目 录

导 言

为什么不是一无所有？

这是一本有关大爆炸（能够想到的最宏大的主题）的小书。它是本有关宇宙学的书，与电视剧（译者注:《生活大爆炸》）无关。关于宇宙学，按照宇宙学家的定义，是将宇宙作为整体，研究它的结构和演化的一门学科。在过去的一个世纪里，宇宙学越来越多地意味着对早期宇宙的研究——研究星系起源、分析质量最轻的化学元素、观测充斥着所有空间的热辐射以及探索那些我们不能直接看到的极端奇异的现象——暗物质和暗能量。总的来说，宇宙学家关心的是我们宇宙诞生后最初的 10 亿年、几年，甚至是不足一秒所发生的事件。宇宙学可以精确地表述为研究宇宙起源的理论——大爆炸。

1

在有些时候宇宙学被称为物理学与哲学的交汇之地，这种说法在很大程度上是正确的，而且是不可避免的。当我们深入研究时，会发现所有科学研究都是在提出问题，追寻问题的答案。如果我们对问题追问得足够深入，那么终究无法摆脱“无解”的命运。在这个难题面前，宇宙学独领风骚。当开展一场有关大爆炸的对话时，任何一位非宇宙学家（大多数人）提出的第一个问题都是“大爆炸之前是什么？”这是个自然而又正确的问题，但目前还没有答案，而且这种无解的状态恐怕要一直持续，直至我生命的尽头。

然而我的计划是摆出这些由外行提出的问题，还有其他的问题，尽量用最简单的方式来回答。因为这本书主要是为那些对科学有好奇心但又缺少科学和数学背景的读者而写，我的同事们会觉得本书既不严谨又不完整，但我的目标并非覆盖尽可能多的领域，而是阐明一小块领域——如果可以的话。

2

为了实现上述目的，我尽量把专业术语的用量降至最低，尽管不得不引入一定数量的公式，并非会让每一位读者都满意，但本书中没有任何一个公式会比描述一条直线的公式更复杂；其他一些复杂的信息，我都把它们扔到脚注里了。我还假设读者能够理解基本的草图，并且愿意读懂非常详细的论证。另外，爱因斯坦有无数从未说过的名言，我同意那句：你应该让事情尽可能地简单，但不要过于简单。过去几年，我已经相信确实存在一定的认知水准，低于这个水准有些事情就不可能简单；对于宇宙学来说，这很大程度源自其天然的数学属性。如果我无法用可以理解的物理概念来解释数学的话，那么我就不去尝试了。

虽然本书不会提及任何涉及真正数学的内容，但其目的之一就是让你相信现代宇宙学是建立在坚实基础之上的恢宏之作，你应该成为其坚定的信仰者。本书的每一章都与前文环环相扣，

因此阅读应该从第一页开始。如果你只对结局感兴趣，那么你就会变得十分没有耐心。　3

正如我之前所说，宇宙学确实提出了深邃的问题。在进行现代大爆炸理论的概念性支撑的探索中，我认为不应该回避这些问题。一位导师曾说:“如果你提出一个愚蠢的问题，你可能会觉得自己很愚蠢。但如果你没有提出一个愚蠢的问题，你依然很愚蠢。”

不可避免地，随着本书内容的深入，问题将会比答案更多。毕竟，当你沉思那些无法衡量的问题时，从最简单的问题“大爆炸之前是什么?”到终极难题“为什么不是一无所有?”之间只有一步之遥。考虑到千年来人类已经以这样或那样的方式不停地追问过这个问题的答案而仍然没有能达成共识，那么也没有理由希望会在本书里找到答案。事实上，如果你向任何一位诚实的宇宙学家提出这个问题，你能得到的回答只有一个“我不知道”。一个更简单的问题是“那些电视节目中白板上的公式有什么意义吗?”，答案是肯定的。个人经验告诉我们，宇宙学家还没有做好回答任何有关化妆品的问题（译者注：宇宙学家“cosmologist”与化妆品“cosmetics”的英文单词有同样的词根，作者说的是反话）。

*

因为本书是为大众而写，所以我宁愿用类比而不是以公式来表达。但由于任何类比迟早都会幻灭，所以这里潜伏着危险。　4

类比，与任何理论一样，是现实的模拟而非现实本身。谈及大爆炸时，宇宙学家通常用气球来解释膨胀宇宙的某些性质，但真实的宇宙不是气球，这样的类比并非完美。当考虑类比时，认清类比和现实之间的区别至关重要。

至此，我已经多次使用“理论”这个词了。我要强调一下，当一位科学家使用到“理论”这个词时，它通常与我们日常生活中的意思是不同的。听众在广播里经常听到某位检察官对某项罪行提出某种理论，而辩方律师的理论认为检察官疯了。通常，这些都是毫无依据的推测，而实际情形又变化多端，以致这些假设理论根本没有意义。

与之相对的是，任何一个物理理论都是理念与猜想的高度交织，既有数学支撑，又被实验和观测证据坚实地支持。当宇宙学家提到大爆炸理论，指的就是猜测与观测织就的理论网，大爆炸理论的各项元素已经被仔细检验了整整一个世纪，有太多精确的观测事实来支持整个理论图景，甚至有些宇宙学家认为宇宙学研究已经更像工程学而非基础研究。所以，相信现代宇宙学吧！

✷

然而宇宙学与其他大多数科学之间还存在一个根本的差别：只有一个可观测宇宙。大多数科学的精要就是实验和重复。例如，制药厂为了检验一款疫苗，会进行多项不同目标的医学实验。如果实验结果不能被全世界其他科学家所复制，那么疫苗的可信度就不会被承认。但至少到目前为止，宇宙学家尚没有在多个宇宙开展实验的机会，因此如果宇宙以不同的方式起源的话，他们无法完全确定地描述整个宇宙会是什么样子。

尽管宇宙学家无法描述所有事物，但他们了解的还是远比空无一物要多得多。仅当我们将宇宙作为一个整体，考虑那些终极问题时，单一宇宙的事实才会对我们造成困难。由于缺乏多个实验对象，宇宙学家就从他们的“表亲”——天文学家那里借用数据和观测结果。天文学家有着通过地基望远镜或近地 6 轨道空间望远镜来研究行星、恒星和星系的传统。当然，天文学家也是一群“旱鸭子”，至今还没有任何一种探测器或者望远镜能够到达离我们最近恒星的周围，更不用说造访其他星系了，而这意味着人类不可能在遥远的天体上进行任何实验。所以，天文学被称为“观测科学”是有道理的。

但是所有的天文学都有一条最基本的假设，就是基本的物理规律在整个宇宙中都是同样起作用的。天体物理学家，他们是宇宙学家和天文学家的近亲，已经用这些物理规律来揭秘恒星和星系的各种行为。因为至少在人类文明的生命期限内，发射空间探测器到宇宙深处是不切实际的，我们只能转而依赖光和其他信使将遥远宇宙的信息传递给我们。实际上，现代科学最伟大的胜利之一，就是假定人类知晓的自然法则在各处都是一样的，即使“足不出户”，我们也已经了解了很多有关宇宙的信息。但是，把已知的物理定律应用到宇宙这个单一对象时，7 这些定律到底在多大程度上有效仍未可知。

宇宙学家试图用与天文学家、天体物理学家同样的办法重建宇宙的演化：用纸笔或计算机，依赖已有的物理规律，以数学自洽的形式构建所研究系统的模型，再核对模型结果与观测数据是否一致。其研究的系统可以是某个星系团或者是宇宙整体。如果我们的模型预测与观测相吻合，那么我们就走出研究室去喝酒庆祝了。反之，就要回头查找模型中是否存在数学错误。假如没发现数学错误，我们再去查找概念错误。如果最终没有一个模型能够与观测吻合，我们就会加入新的现象。如果新的现象改进了结果，我们就会请负责观测的同事开始探测。

任何一位科学家最不愿意做的事，就是在穷尽各种平凡的解释之前，在现有模型中加入奇奇怪怪的新现象。但当考虑宇宙大爆炸之后极早期的瞬间时，就呵呵了……

8

✷

此刻，你可能会对天文学家和天体物理学家转身离开，交由宇宙学家接手的精确时刻感到困惑。这里没有明确的边界，通常研究其中一面的科学家，也了解很多另一面的信息。最重要的差别就是“尺度”。如前所述，传统而言，天文学家和天体物理学家关心的是恒星、行星以及星系的行为，最近他们的研究对象也扩展至整个星系团甚至超星系团——星系团的团。而宇宙学家关心的则是能够想象到的最大图景，起始于超星系团的尺度，试图了解所有这些结构是如何呈现出我们所观测到的宇宙的样貌的。虽然主导星系行为的物理规律与作用于恒星和行星的完全一致，但本书并不准备讨论这些。可能稍稍涉及的话题是黑洞，它们总是那样引人入胜。从宇宙学家的角度来说，以上这些天体太小了，都不重要。

对宇宙学家来说，头脑中总是明确各种天文尺度是极其重要的。本书中我都会采用天文学术语来表述距离，也就是光在一定时间内穿行的距离。你大概知道光需要大约 8 分钟的时间才能从太阳传至地球。简单起见，我们就算是 10 分钟。这样我们就可以说，地球处在距离太阳 10 光分的地方。同样的，1 光年就是光在 1 年的时间里走过的距离。天文学家从来不会把光年换算为英里或者千米，你也不要那样做。事实上，你应该培养一种对宇宙中不同尺度的直觉：　9

4 光年——太阳之外最近恒星的距离。

我们的银河系直径大约是 10 万光年。

星系团的直径大约为几百万光年。

超星系团的直径大约为几亿光年。

可观测宇宙的大小约为 140 亿光年。

✷

这就是宇宙的尺度，也是本书要谈论的尺度范围。

你能给我有关眼影和睫毛膏的建议吗？不能！　10

第一章

引力、南瓜和宇宙学

宇宙学是研究引力如何决定整个宇宙演化的学科，因此要理解宇宙学必须搞清楚引力。

引力是目前已知自然力中强度最弱的一种。对物理学家来说，力只不过是施加于某物体的推力或拉力，再无其他——这里不存在“不明地带”——这也是物理学家把自己的研究领域视为全部科学的最基础部分的主要原因之一。在过去的几个世纪里，物理学家已经了解到自然界只有四种基本作用力。第一种被称为“强核力”，顾名思义，它是最强的一种自然力，将原子核的各成分束缚在一起。任何一种原子核都由中子和质子构成（译者注：氢原子核除外，它只包含一个质子），带正电的质子之间存在着相互排斥的电作用力，如果没有强核力把原子核束缚住，那么质子间的电斥力就会让原子核分崩离析。与强核力相关的能量就是原子弹爆炸时产生的能量释放。虽然强度大，但强核力的作用范围仅限于原子核内部，就宇宙学而言，那实在是太小了。

11

第二种基本作用力是“弱核力”。作用强度只有强核力的几十亿分之一，弱核力主导着一定形式的辐射衰变。氚，也称超重氢，是氢的一种同位素，具有放射性，能衰变为氦；氚的衰变率就由弱核力决定。但与强核力一样，弱核力的作用范围也仅限于原子核，从宇宙学的角度来看，微不足道。

日常生活中，最重要的力是电力和磁力，二者其实是名为“电磁力”的基本作用力的一体两面。全部化学现象都起因于电磁力，任何需要电流的设备都涉及电磁作用，从烤箱到智能手机，乃至今天我们认为理所当然的万事万物。电磁力是现代文明的基础。但为了产生电力或者磁力，就必须有电荷。由于各类天体，例如行星，都是电中性的，因此天体之间并不存在电磁力相互作用。

12

所有物体之间都存在相互吸引的引力作用。然而引力又弱得难以想象——整个地球所形成的引力拖曳都无法让冰箱磁贴稍稍挪动一下，这就是与电磁力相比，引力有多弱的最好实证。而物理学家则试图这样阐明引力作用的强度：两个氢原子核（即质子）之间的引力作用，大约要比它们之间的电斥力弱 36 个数量级。在设计家用电子产品时，工程师们完全不会在意引力。

正是由于核力只作用于原子核内部，而天体都是电中性的，所以在决定宇宙命运的作用力中，自然界只剩下强度最弱的引力了。

✵

现代引力理论即阿尔伯特·爱因斯坦的广义相对论，被称为最美的科学理论。这是真的。 13

流于表面的话，我们可能会肤浅地认为广义相对论只是对近 400 年前由伊萨克·牛顿提出的引力理论的改良。牛顿引力理论仅用一个不朽的公式就描述了引力是如何作用于两个物体的，力的大小取决于二者的质量和间距。我们甚至不需要写下公式就可以理解其传递的信息：只要知道两个物体的质量和它们之间的距离，就可以精确计算出它们施于彼此的引力作用。①

正如前文所述，在物理上，力可简单认为是推或拉。更准确地说，为使物体改变运动速度，也就是使其加速。如果一架钢琴的运动在变快或变慢，说明有一个力施加其上；如果它保持恒定速度，则说明没有力施加其上。 14

按照牛顿理论，如果我们知道施加于一个物体的力，我们就知道它的加速度，从而能够完全预测它未来的轨迹。因此，如果知道宇宙中所有恒星的质量以及它们现在的距离，我们就可以了解有关宇宙未来的一切——当然也包括它的过去。正因如此，牛顿的宇宙通常被类比于时钟。就大部分而言，确实如此。

---

① 举例来说，牛顿定律给出质量分别为 m_1 和 m_2 的两个物体间引力 F 的计算方法，$F=Gm_1m_2/r^2$，其中 r 是二者的距离；G 是“引力常数”，一个必须在实验室中测量的数值，决定了引力的强度。

✷

牛顿的引力理论与日常生活吻合得如此之好，以致在整整200年时间里，天文学家认为它完美地解释了太阳系的所有运动。直到19世纪中叶，第一条显示其并非完美的线索才隐隐现身。与其他行星一样，水星也在绕太阳的椭圆轨道上公转。如果太阳系只有水星和太阳，那么每一次水星最接近太阳的位置——水星近日点——就会是太空中固定不变的一点。但天文学家观测到的却是水星近日点的位置一直随时间在改变。计算结果表明，太阳系其他行星的引力拖曳作用能够解释水星近日点漂移的绝大部分，但仍顽固地残留下无法解释的一小部分。虽然提出很多理论来解释这一小部分移动，但理论计算与观测之间的差异好似幽灵一样存在了半个多世纪。

20世纪初，爱因斯坦开始研究广义相对论，当时除了水星近日点漂移之外，没有其他观测证据说明牛顿引力理论有可能是不完备的。然而，詹姆斯·克拉克·麦克斯韦有关电磁领域的理论横空出世了。

首先，你应该认识到牛顿理论是有关粒子和力的理论。就像躺在南瓜地里的两个南瓜，我们可以把它们理解为两个粒子，彼此施加了穿过南瓜地的引力作用。类似地，我们也可以把地球和月球抽象为两个粒子，彼此施加了跨越太空的引力作用。但两种情况里，牛顿理论都没有解释引力是如何从一个粒子传递到另一个粒子的。因此，牛顿引力作用通常被称为一种“超距作用”理论，在牛顿时代，“作用”一词意为“力”。

同样重要的是，显而易见，两个物体间的引力作用是瞬时发生的；如果太阳突然消失，行星就没有了绕其公转的中心天 16 体，会立即逃逸到太空，一刻都不迟疑。

✷

抛开刚才提到的南瓜地，想象一下漂浮在池塘中的南瓜。我们立刻就能意识到图景发生了变化。池塘中的水是由无数分子组成的，但水分子太小了，我们忘记分子转而把池塘中的水视为在每一点具有特定密度和压力的物体。密度和压力是“宏观”量，并不考虑单个粒子。这是一个典型的“场”，一个房间中的空气可以被当作一种“场”。蹦床的弹性表面也可以被当作“场”，蜂群在很多方面也与场相似。

场的画面为引力传播提供了一种非常自然的机制。如果池塘中的南瓜上下浮动，它们会产生小小的波动，这些波动会随着水波传播到整个池塘。这些波动是产生于局部的扰动，通过水形成的场，以有限的速度传播开来。作为对比，在牛顿引力理论中，人们需要想象存在能够以无限速度、以某种方式穿过巨大空间的力。 17

“反对！”你可能会礼貌地叫嚷：地球和月球间的引力作用并不涉及波。确实如此。所有的类比都会瓦解。当我们考虑两个物体间永恒的引力相互作用时，到底把它想象为力还是场都无关紧要。然而，场确实存在；如果你曾把铁屑撒在一张纸上，然后在纸面下移动一块磁铁，你就会对磁场的形状有非常直观的感知。整体而言，场的图景是如此强大，几乎所有的现代基础物理理论从本质上说都是场论。如果没有场的概念，我们很难想象该如何描述电磁波以及引力波。

可以肯定的是，当麦克斯韦思考主导电场和磁场的定律时，他已经指出这些场能够以电磁波的形式穿过宇宙空间，速度恰好是 3.0×10^8 米每秒。[①] 麦克斯韦的发现发表于 1865 年，他本人也震惊于这个结果，因为这个数值几乎精确等于光速（在当时已经被精确测量过）。其结论就是“几乎不可避免的”，他写道，“光本身”一定也是一种电磁波，并不是以无限的速度传播，而是以有限的每秒 3.0×10^8 米的速度传播。麦克斯韦的预言是 19 世纪物理学最伟大的理论胜利，几十年后发现的射电波再次证明了它的正确。

18

20 世纪开端，很多物理学家试图以麦克斯韦的电磁理论为基础，创造引力场论。但他们都无一例外地失败了，因为引力的行为与电磁学并不完全相似。爱因斯坦是第一位理解二者区别的物理学家，也首先将引力理论拉上正轨。为了理解爱因斯坦描述引力场的理论——广义相对论——我们必须先对他早些时候提出的狭义相对论有充分的认知，以此为起点才能明白广

① 科学记数法是物理学和天文学不可或缺的。向不熟悉科学记数法的人解释一下：指数表示的是数字 10 的幂，或者说是数字 1 后面有多少个 0。因此，数字 10 可以写为 10^1，100 可以写为 10^2，1000 可以写为 10^3。3×10^8 就是 300000000，这就说明了为什么我们要用科学记数法。

义相对论到底是怎么回事。

什么是相对的？什么又不是相对的？　19

第二章

狭义相对论

自 19 世纪 20 年代以来，自然哲学家们就已经知道电和磁是紧密相关的。电流能够产生磁场；反之亦然。麦克斯韦的电磁理论精确地描述了电和磁到底是如何互相产生的。爱因斯坦在提出狭义相对论时指出，电和磁不仅仅是紧密相关的，实际上二者是同一自然现象的两个方面。正因如此，爱因斯坦发现牛顿物理学必须被修正。

20

但是爱因斯坦从未同意过那句名言“一切都是相对的”。说到底，所有的物理学考虑的都是运动，而相对论要回答的本质问题是：某物的运动状态改变时，究竟有什么发生了变化，又有什么是保持不变的？有些事物发生了变化，同时另一些则保持不变，因此相对论理论或许被称为“绝对理论”更精确，事实上当初它就是这样被提出的。

相对论中最重要的绝对量就是光速。麦克斯韦的发现里最奇怪的事情就是，电磁波在真空中传播的速度约为每秒 3.0×10^8 米，这个数值现在通常以字母 c 表示，它自然而然地就出现在麦克斯韦的方程组中。当我们测量一列火车或者一个棒球的速度时，总是以某个其他物体作为参照。如果我们站在田野中，可能会测量到一列朝东行驶的火车相对于大地的速度是每小时100千米。然而，当我们坐在一辆同样朝东，行驶在与火车轨道平行公路上的汽车里时，如果汽车的行驶速度为每小时75千米，那么从汽车里看起来，火车的速度就只有每小时25千米了。我们测量到的任何物体的速度，总是依赖于我们的参考系——粗略地说就是我们的视角，更具体地说就是我们站在哪里。[21]

麦克斯韦的结果之所以奇怪，是因为它只表明光速 c 的大小是每秒 3.0×10^8 米。这个速度是相对什么的？麦克斯韦自己认为他的电磁波可能穿过了名为“以太”的发光物质。

水波通过水来传递，声波通过空气传递，因此很自然地会认为光波也必须通过某种媒介来传递。发光的以太应该充满整个空间，并提供了一个绝对静止的参照物。如果你坐在一列火车里，那么相对于火车你就处于静止状态，但相对于地球，火车则处于运动状态，而地球又相对于以太在运动。同样，水星也有相对于以太的运动速度，这样你就可以比较水星和地球的速度，因为它们各自都有相对于以太的“绝对速度”。麦克斯韦认为光相对于以太的绝对速度就是每秒 3.0×10^8 米。

不幸的是，简单计算的结果表明以太有着非常诡异的性质。例如，如果以太要比空气稀薄100倍，那么它的硬度就得比钻石大1000倍。说得更直白一点，所有试图探测以太的努力均以失败告终。[22]

✷

1905年，爱因斯坦终于极其大胆地提出以太根本就不存在。更进一步，他接受了麦克斯韦光速为常数 c 的结论，并使其成为自然界的一条定律。爱因斯坦的狭义相对论就这样横空出世了。它基于两个基本假设。

第一，不存在绝对运动。爱因斯坦从伽利略那里借鉴了这个公理，没有任何在一列火车上进行的实验能够区分火车到底是处于静止还是在匀速运动。所有的运动都是相对于某个参考系来测量的，没有哪个参考系更优于其他参考系。

第二，处于任何参考系的任何观测者测量到的真空中的光速都是常数 c，大小为每秒 3.0×10^8 米。

这里我们需要几条犹太法典式详细的注释。第一个假设称为“相对性原理”（爱因斯坦最初并没有把自己的理论称为相对论；这个名称是后来加上的，最初的理论名为“绝对理论”）。之所以被称为“狭义”，是因为该理论讨论的是匀速运动。爱因斯坦并没有讨论加速运动，并且假定上述的所有参考系本身也都是匀速运动的。在相对论中，运动才是真正相对的。

23

第二条假设看起来简单，但它改变了一切。认为处于任何参考系的任何观测者都会测量到同样的光速，这与牛顿物理学直接矛盾。如果光好像火车穿过高速公路一样运动，那么其速度应该依赖于观测者（按照物理学家的习惯，进行测量的任何人或物都被称为观测者）的参考系。

*

光速为常数的假设还表明时间和空间不再是几个世纪以来人们一直认为的那样完全割裂的。很容易就能搞清楚为什么会这样。让我们想象有一个时钟，它由在一列压扁的火车里上下弹跳的小球构成（见图 1 上）。

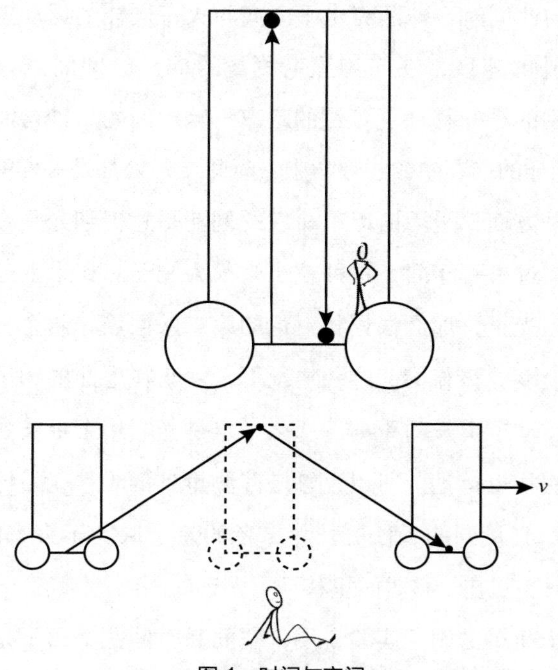

图 1　时间与空间

火车上的鲍里斯看到小球只是直上直下地往返运动，他可以将 1 秒定义为小球从火车地板到天花板间一次往返运动所需要的时间。

然而如图 1 所示，娜塔莎是在地面观察火车，她看到火车朝右侧以速度 v 在运动。对她来说，1 秒仍然是球完成一次从地 25 板到天花板往返运动的时间，但相对于地面来说，小球是沿着一个三角形运动的，因此移动的距离更大。

不仅如此，娜塔莎看到的小球运动速度也更快。在垂直方向，小球的运动速度与鲍里斯看到的是一样的，但对于娜塔莎来说，与此同时小球还有水平方向与火车同样的运动速度。由于这额外的速度，在与鲍里斯测量的同样时间间隔里，娜塔莎测量到的小球移动了更长的距离，鲍里斯的 1 秒与娜塔莎的 1 秒是同样的。在牛顿物理学中，时间在各处都是一样的。

另一方面，爱因斯坦革命性的创新是意识到光是由粒子构成的，在过去百年间，这种粒子被称为光子。如果上述火车里的小球是光子，那么根据相对论的第二条假设，两个观测者测量到的速度应该相同。这种情况下，由于在地面的娜塔莎看到的光子运动了更长的距离，那么一定是光子用了更长的时间来完成一次往返运动。所以，娜塔莎测量到的 1 秒就要比火车上的鲍里斯测量到的 1 秒更长。二者的差别取决于火车的速度，也就是取决于在 1 秒时间里移动了多远的距离。

这个简单的想象实验说明，空间和时间测量再不是彼此独 26 立无关的。爱因斯坦精确地描述了二者是如何相关的，但就我们的目标而言，那些细节就无需在此赘述了。由于相对论的出现，物理学家不再分别考虑空间和时间；取而代之的是他们所谈论的四维“时空”，它将空间距离与时间联系起来。

虽然时空的概念隐含在狭义相对论中，但爱因斯坦并不是首创人。他早期的有关相对性的论文中，没有一处将时间认为是时空的第四个维度。法国数学家亨利·庞加莱更早看出时空概念的必要性，而德国数学家赫尔曼·闵可夫斯基是第一个给出其数学表达的人。爱因斯坦甚至反对时空的概念，认为它是“多余的认知”。然而最终证明，时空的观念对于广义相对论的建立是至关重要的。

✷

狭义相对论还有其他革命性的结论。其一是光速给出了速度的上限，没有任何观测者能够测量到有质量物体的运动速度会比光速更快。其二是随着有质量物体速度加快，它的质量也随之增大，当达到光速 c 时，其质量也会变为无穷大（这也是没有任何有质量物体的速度能超过光速的原因之一）。

27

当然还有另一个革命性的结论，就是重要的爱因斯坦质能方程 $E=mc^2$，即物体固有的能量等于它的质量乘以光速的平方。然而根据定义，光每年走过 1 光年的距离，因此在这个量纲系统里，光速 $c=1$，所以质能方程简化为 $E=m$。自相对论提出之时起，物理学家就把能量和质量视为同一事物的两面，所以他们认为“质量密度”和“能量密度”是等价可交换的，本人也这样认为。

与通常认为的相反，爱因斯坦并不是第一个提出质量和能量相关的人，虽然这么说会显得很不敬，但爱因斯坦确实从没有成功地证明过质能方程 $E=mc^2$。他有关于此的著名论文中有一个错误，虽曾试图随后修补，但没有成功。尽管如此，从其在解释原子弹爆炸或者太阳内部核反应中所扮演的核心角色来看，质能方程的结果经受住了时间的检验。

28 还有什么是狭义相对论没有谈及的?

第三章

广义相对论——宇宙学的基础

现代宇宙学本质上来说，就是把爱因斯坦的广义相对论应用于宇宙整体。如今，广义相对论至少已经成为历史上被证实的最精确科学理论之一。任何已经进行的实验或者观测结果都不违背广义相对论，宇宙学家也不再有任何质疑，完全相信广义相对论已经为我们的宇宙提供了完美的描述。

虽然广义相对论涉及的数学非常复杂，但它的基本要义还是能够理解的。在把话题转向宇宙之前，我们应该试着搞清楚为何一个被称为广义相对论的理论成了一个有关引力的理论，我们为什么会相信它是正确的，它又是如何以独特的视角塑造了我们的时空观。

29

如果说几乎所有的物理学都是关于运动的，那么在前述的篇章里我们就忽略了最基础的事情：加速度，也就是速度的改变量。在提出狭义相对论时，爱因斯坦假定物体都是匀速运动的。没有任何物体被加速，因为没有力就不会产生加速，因此狭义相对论的框架里也没有力①。

爱因斯坦打算扩展狭义相对论，要把加速考虑进来——于是他进而提出了广义相对论。如果说广义相对论总是被称为最漂亮的理论（确实如此），那是因为虽然描述广义相对论的方程十分复杂，但整个广义相对论大厦及其全部预言，都是建立在两个非常简单却坚实的假设之上。

✷

让我们从爱因斯坦自称为“一生中最幸运的想法”开始。30 自伽利略时代起，我们就已经知道如果没有空气阻力，那么所有物体都会以同样的速度坠落地面。这就是著名的引力加速度，通常用字母 g 表示。在地球表面附近，g 值刚好是 9.8 米 / 秒 ^{2}，但对于我们来说，具体的数值并不重要，除非你是工程师。对物理学家来说，重要的是 g 值与坠落物体的质量和组成都无关。金块也好，西瓜也罢，包括羽毛在内的所有物体，在真空中坠落的速度都是相同的。

同样的原因，如果我们乘坐的电梯被切断电缆，我们会突然感受到失重，因为此时我们与电梯正以同样的加速 g 在坠落，脚底不再对电梯地板施力，或者换成浴缸里的场景，我们已经方便地体会过失重。

在小局域范围内，自由落体和引力消失这两种状态是无法

① 经过某种操作，加速和力能够从形式上被融进狭义相对论，但这样做并不能直接转换为广义相对论。

区分的。

国际空间站里的情形正是如此：宇航员与空间站以同样的速度环绕地球运动，因此就感受到失重。一个更常见的体验是，当我们乘坐的电梯向上运动时，我们会感觉比平时更重。此时，引力似乎变大了。

31

爱因斯坦把这些简单的观测提升为自然法则，称为“等效原理”：

> 在一个足够小的封闭环境中，没有任何实验能够区分匀加速运动和均匀的引力场。

换句话说，如果电梯没有窗户，处于电梯中的人是无法分辨到底是电梯在加速上升，还是地球的质量突然增大导致其引力场增强（“引力场”是引力产生的加速度 g 的另一种表述）。同样，如果电梯的电缆受损，我们也无法分辨到底是自己正以加速度 g 在朝地面坠落，还是地球突然消失了。对于局部来说，加速度和引力场是等效的。

正因如此，爱因斯坦意识到要把加速度融入狭义相对论，必须有一种关于引力的全新的理论。

*

爱因斯坦关于引力的理论，尽管一直被冠以容易被误解的名字——广义相对论，但要比他的狭义相对论更深远地改变了我们对时空的认知。单单是等效原理，就要求同处于地球引力场中不同海拔处的时钟，必须以不同的频率计时。这不仅是每

32

天都要发生几百万次的事实，而且如果不是这样，那么现代生活中的大部分都不可能实现。

让我们把爱因斯坦曾做过的一个思维实验稍稍升级一下，想象一艘加速向上驶向空旷太空的飞船。为了便于通信，位于飞船顶部的娜塔莎手里有一部手机，位于飞船尾部的鲍里斯手里也拿着一部同样的手机。娜塔莎按照自己手机的时钟，每秒都用手机的“等效应用程序”向鲍里斯发出一个光信号。但由于鲍里斯在光信号传播的时间内，正在做向上的加速运动，因此此刻他的速度已经比光信号发出时更快了，所以他会比做匀速运动更早收到娜塔莎发出的光信号。鲍里斯看到的连续两个光信号的时间间隔要比娜塔莎看到的短，因此他就得出结论，自己的时钟比娜塔莎的运行更快。[①] 如果加速度和引力场是等效的，那么同样的事情也一定会在地球的引力场发生。

33

全球定位系统（GPS）就是依靠位于地球上方的在轨卫星提供的时间信号来实现导航的。因为卫星绕地球高速运动，按照狭义相对论，卫星上的时钟要比地面手机的时钟运转得更慢。但由于高轨卫星轨道所处的引力场更弱，按照广义相对论，卫星上的时钟应该更快。根据广义相对论得出的时钟差异刚好是狭义相对论计算结果的 2 倍，但这种差异总计也不超过每秒十亿分之一秒。

在十亿分之一秒的时间里，光传播的距离大概是三分之一

---

① 有些读者可能已经明白我正在描述的是多普勒位移。

米，也就是 1 英尺，因为光速是每秒 3.0×10^8 米。除非 GPS 系统按照相对论修正误差，否则你的 GPS 定位每秒就会偏离大约 1 英尺。那么几分钟之后，那些不再懂得看地图的人就会不可避免地迷路。

广义相对论是正确的。 34

*

广义相对论对宇宙的描述是牛顿不曾意识到的。你很可能知道源自伽利略的著名的牛顿惯性定律——任何物体都趋向于保持原有的运动状态。更准确地说，如果没有外力作用于物体，那么它就沿直线运动。引力使得物体沿曲线运动，正如你抛出小球后，小球沿曲线运动最终落到地面那样。但正像我们刚刚看到的，在自由落体的电梯里，引力消失了。所以在这部电梯里，不再有力作用于小球，根据惯性定律，小球必然沿直线运动，如图 2 左图所示。

爱因斯坦认为光也遵循同样的定律。所以，在一部自由下落或是匀速运动的电梯里，如果没有外力作用，光就会沿直线传播——仍如图 2 所示。但如果电梯以加速度 g 向上加速运动，或者电梯位于引力场大小为 g 的行星上方，那么等效原理就要求光必须偏折，两种情况下偏转的量相同，如图 2 中图和右图所示。

太奇怪了：按照前面章节所述，似乎物体沿直线还是沿曲线运动，依赖于选择的参考系。更奇怪的是，似乎引力是否存 35

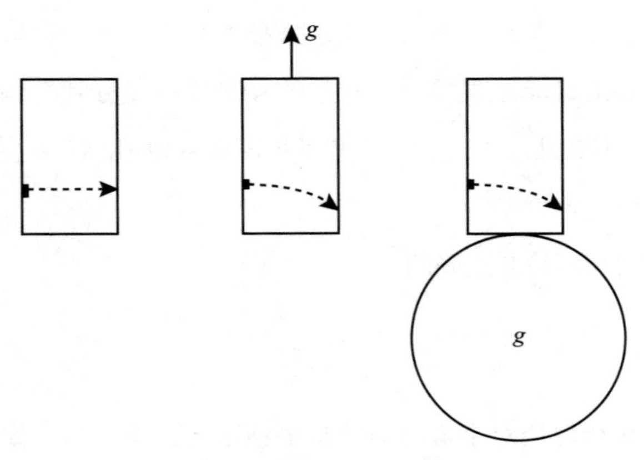

图 2　电梯中小球的运动

在也依赖于参考系。但这是真的。

让我们想象未来能够建造一座大厦，其高度与地球直径相比不能忽略。地球引力加速度 g 的值，在大厦的顶部测量到的比在地面测得的值要小。这样的情形就不再是前述的“小局域环境”了。

如果分别位于靠近大厦顶部和底部的两部电梯都突然停电，它们将以不同的加速度下落。让两部电梯内的人都同样扔出小球，他们自己都看到小球是沿直线运动的，但对于能同时看到两部电梯中小球运动的人来说，他会看到两个小球是沿不同曲线运动的。这种情况如图 3 中图所示。作为对比，图 3 左图所示的是在更矮建筑内两部电梯里抛球的情景，由于加速度 g 保持常数，因此两个小球的运动轨迹是相同的，决不会交叉。如果那栋摩天大厦侧向倒下，两个小球都会朝地球中心坠落，最终它们的运动轨迹将会交汇，如图 3 右图所示。

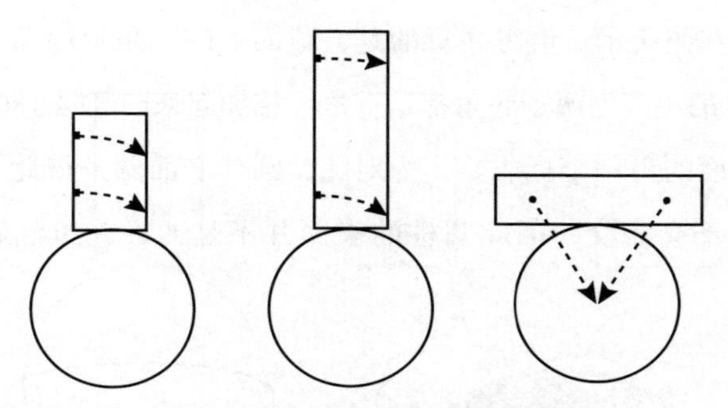

图 3　不同高度的电梯中小球的运动

相隔较近的粒子拥有相同的运动轨迹，而相隔较远的粒子则有不同的轨迹，这正是一种潮汐现象。对于地球来说，距离太阳更近的一面所处的引力场，要比相对的另一面更强。两面受力的不同就导致对地球的拉伸，造成了著名的潮汐隆起和海洋潮汐。

37

正如已经提及的，我们总能找到一部处于失重状态的小电梯。当我们以全球视角来观察时，就会体会到潮汐，而在地球表面，无论我们以何样的视角来看，潮汐都不会消失。在牛顿力学中，潮汐是引力最清晰的表现。

现代宇宙学家用几何语言来描述引力。在一张平铺的纸上，两条平行线永远不会相交。事实上，这正是欧几里得几何著名的第五条公理。在狭义相对论中，没有任何力的作用，粒子沿着平行轨迹运动，并一直保持。狭义相对论是关于平直空间的理论。

然而，在弯曲表面，两条最初平行的线可能最终会相交。例如，在地球赤道处平行的两条经线，会在南极或北极相交，

38 如图4左图所示。值得注意的是，球面上的三角形内角和是超过180°的（因为两个底角都是直角，相加起来已经是180°了）。这是曲面的另一个标志。作为对比，圆柱表面两条彼此平行的直线永不会相交，因此圆柱的表面并不是曲面，虽然看起来如此。

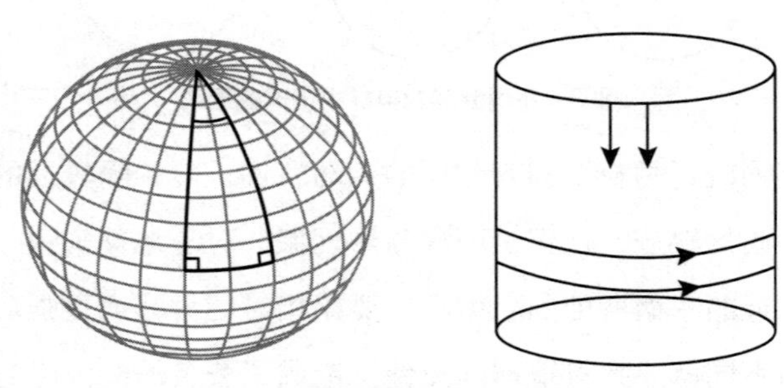

图4 不同表面上的平行线

引力造成的情形与此相同。在电梯里，粒子的运动轨迹彼此平行，但距离更远的粒子运动轨迹则遵循弯曲表面的特征，初始平行的轨迹最终会相交。有些物理学家认为相对论的几何图景已经是一种与物理研究无关的类比了。但是广义相对论的几何特性，延展至包括第四维度的时间在内时，的确就是曲面几何，这是由乔治·伯纳德·黎曼以及19世纪的其他数学家提出的。39 如果认为它是种类比，那么它的确是个完美的类比。引力就是弯曲空间——也就是时空。

牛顿的引力理论告诉我们有质量的物体就会产生引力，引力会让其他物体运动。广义相对论则告诉我们，质量会弯曲时空，这种弯曲将决定其他质量如何运动。如果说牛顿的宇宙中力是作用于永远平直的空间，那么在爱因斯坦的宇宙里，随着质量在其中移动，空间时间都一直在改变形状。这就是广义相对论革命性的概念。

1915年，爱因斯坦完成了广义相对论，能够准确解释水星近日点进动问题；由于水星是太阳系最内侧的行星，那里的时空弯曲显著，与牛顿引力体系的差别已经达到可以被测量的程度。由阿瑟·爱丁顿领导的1919年日全食观测表明，经过太阳附近的恒星光也会被太阳的引力场弯折，正合爱因斯坦的预言。一个世纪之后，广义相对论已经成为史上被最精确检验的理论之一。如今，查看地图已经成为失落的艺术，就是广义相对论活生生的实证。

40

*

与电磁理论一样，广义相对论也是场论，允许波的传播。在第一章中我们已经知道，广义相对论并不是第一个引力场论，爱因斯坦也不是预言引力波的第一人。实际上，爱因斯坦最初并不相信引力波，甚至在意识到引力波存在后，他的第一篇相关主题的论文完全是错误的。但爱因斯坦的确是正确理解引力波的第一人。

在电磁理论中，加速运动的电荷会产生电磁波——可见光波或者射电波——同样，在广义相对论中，加速运动的质量会产生以光速传播的引力波。但引力波不是光波，无法用普通望远镜探测到。事实上，引力波是穿过时空的极细小的潮汐波动，会拉伸或压缩探测器本身，正如月球潮汐拉伸压缩地球那样。由于引力波太弱，所以探测引力波的难度超乎想象，其对探测器的拉伸量，比质子直径的万分之一还小。但是，经过半个世41纪的努力，研究者终于完成了这个奇迹。2016年，激光干涉仪引力波天文台（LIGO）宣布探测到引力波。探测到的引力波由位于10亿光年之外的两个黑洞碰撞产生，与广义相对论的预言精确吻合，引力波探测成功开创了天文学的新纪元，有些宇宙学家甚至激动得泪流满面。

✷

结果如大家所知道的，广义相对论的正确性达到了科学理论的顶峰。物理学家称其为“经典”理论，意思是它未考虑量子力学。也许很有必要创造量子引力理论，用来描述大爆炸奇点，这是个随后要被反复提到的话题。除了奇点这种极端情况，广义相对论适用于任何能想到的环境，有鉴于此，宇宙学家毫不犹豫地用广义相对论描述整个宇宙的演化。

正如我们将要看到的，真实的宇宙几乎是平直的，或者说是欧几里得式的，因此对于现代宇宙学来说，大多数严谨的广42义相对论理论近乎多余；通常情况下，牛顿理论就足够了。然而，相对论的观点仍是至关重要的。在黑洞这样的天体周围，引力场极强，时空不再是平直的，这种情况下就必须让广义相对论发挥作用了。

✷

至此我还没有谈论过广义相对论的第二条假设。它的名字过于怪异，所以不妨称为“广义”的相对性原理。狭义相对论是讨论匀速运动的理论——更精确地说，是匀速运动的参考系——爱因斯坦认为所有这样的参考系都是等效的。没有哪一个参考系能代表绝对空间。提出广义相对论时，爱因斯坦宣称我们必须能够在任何参考系——特别是加速运动的参考系中来描述运动。

这就提出了非常深刻的问题。

很多人都去过游乐园，玩过绕中心旋转类似离心机的游戏。实际上，我们会说是离心力把我们推挤在离心机轿厢壁上，我们的感觉的确是这样。但一位站在地面的反对者会说，不是这样的，那只是我们的想象所虚构出来的。如果轿厢突然消失，从地面上看，我们会沿着直线飞出去，这符合牛顿的惯性定律。我们感觉到的离心力是“虚构”的。事实上，轿厢壁在向“内”推挤我们，免得我们飞到空中去。

43

旋转的离心机游戏代表着加速运动的参考系，按照很多入门教科书的说法，物理学在这种条件下不再成立。离心力之所以被视为虚构的，是因为从静止的地面看，它就不存在了。但我们已经看到在下落的电梯中，引力自身是如何消失的，它等效于一个不加速的参考系。那么，引力也是虚构的吗？

这个问题的答案是：如果我们相信广义相对论，那么我们就没有别的选项，只能相信要么引力是种虚构的力，要么这个“虚构”的力是真实存在的。

✷

而这又进一步提出了更深刻的问题。当我们坐在火车里时，[44]根据狭义相对论，我们无法辨别火车是匀速运动还是静止，但一定会知道火车开始加速——我们会被向后推向座椅。

那么火车是相对“什么”在加速呢？牛顿认为是相对于绝对空间——也就是永远保持静止的以太。物理学入门级的教科书会赞同牛顿的观点，并认为以太确实存在。

在提出广义相对论时，爱因斯坦深受德国物理学家和哲学家恩斯特·马赫的影响，他认为绝对空间只是牛顿假想出来的。因为无法探测到绝对空间，那么只能考虑加速是相对于其他物体——例如，恒星。爱因斯坦把这个想法命名为“马赫原理”。

1851 年，利昂·傅科将一个非常长的单摆挂在法国巴黎先贤祠的穹顶，这是对马赫提出的难题最著名的展示。随着一日时光的流逝，看起来单摆摆动的方向相对于先贤祠的地板缓慢[45]发生变化。实际上是先贤祠在绕着单摆转动，而后者相对于头顶的恒星以同样的方向转动。那么傅科的单摆是如何“知道”要相对于恒星摆动呢？又或者是恒星构成的参考系刚好与绝对空间相一致？有些人可能甚至没有看出这其中的问题。而其他人则看到了物理学中最深刻的秘密之一。

爱因斯坦试图将马赫原理纳入广义相对论。在一个空无一物的宇宙中，应该根本探测不到任何加速运动。爱因斯坦的努力究竟获得多大成功仍有争议，但对这个问题的探讨应该是另一本书的内容了。所以我就此打住。

相对论是如何描述整个宇宙的？　46

第四章

膨胀的宇宙

今天，宇宙在膨胀的想法已经如此深入人心，成为我们流行文化的一部分，但它到底意味着什么？任何一场关于宇宙学的讲座之后，来到讲台的观众提出的第一个问题都是："如果所有的星系都在远离我们，那是不是意味着我们处于宇宙的中心？"第二个问题则是："宇宙究竟会膨胀到哪里去？"说实话，有些时候这两个问题的顺序会互换，但由于它们被理所当然地提出来，恰恰说明膨胀宇宙的概念并不易被理解。

47 当然，我们不能将这一点归因于爱因斯坦。1916 年，当他发表广义相对论时，还没有任何天文学的证据证明宇宙正在膨胀，因此同年，爱因斯坦用广义相对论来建立第一个现代宇宙学模型时，假定宇宙是静止的。在随后的几十年间，天文学家不断被推向膨胀宇宙的观点，因为他们发现那些星云——通常被认为是银河内部"云团"的结构——其实是位于银河系之外；而且它们看起来正在远离我们。直到 1929 年，当埃德温·哈勃提出他著名的"定律"，宣称遥远星系的退行速度与其距离成正比，我们才彻底接受了膨胀宇宙的概念。由于某些将来也许会更加明晰的原因，哈勃定律意味着星系不仅是在远离银河系，而是在彼此远离。[①]

这就是天文学家在说到膨胀宇宙时的准确含义——星系是在彼此远离。这是宇宙学最为重要的发现，是整个大爆炸理论的基础。显而易见，如果宇宙不是正在膨胀，那么也就不存在大爆炸。 48

✷

从概念上讲，哈勃所做的工作其实非常简单：他只是把一些星系离我们的距离及其速度画在了一张图上。根据哈勃当时的观测数据画出来就像图 5 这样，他要么是极有勇气，要么是愚蠢蛮干，在图上画了一条穿过这些数据点的直线。

我保证，现在我们必须面对本书中最难的数学部分：一条直线的方程。哈勃画出的那条直线对应的方程是 $v=Hd$，其中 v 代表星系的速度，d 代表星系的距离，H 是直线的斜率。直线意味着星系的退行速度与它的距离成正比。如果星系 B 的距离是星系 A 的两倍，那么星系 B 的退行速度（远离我们的速度）也是星系 A 的两倍。而斜率 H 的值越大，说明给定距离处星系的退行速度就越快。 49

① 近来，哈勃定律被更名为哈勃 – 勒梅特定律，增加了比利时教士乔治 · 勒梅特的名字，他于 1927 年提出了相同的定律，只不过是用法语发表的论文。

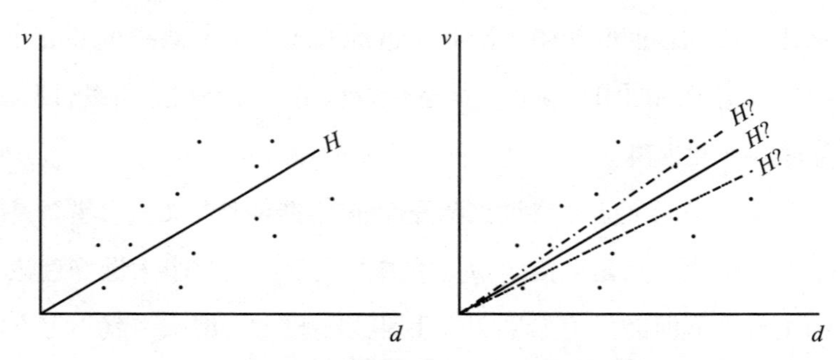

图 5　星系的速度与它们到地球距离的关系

斜率 H 被称为哈勃常数，显然是宇宙学最有名的数值，很多宇宙学家整个职业生涯的目标就是确定 H 的精确数值。为什么 H 值这么重要呢？知道了 H 值也不大可能影响竞选结果，但正如我们将要看到的，H 值测量的是宇宙膨胀到底有多快，而这个结果最终会以某种方式影响所有的宇宙进程。而且知道了 H 的值，我们就知道了宇宙的年龄，也就是自大爆炸起经历的时间长度。理论上说，H 值很容易确定：根据哈勃定律，只要把星系的速度和距离绘制在一张图上，读出那条直线的斜率就可以了。那句老话“说起来容易做起来难”，说的就是测定 H 值这件事。

按照著名的多普勒位移定律，相对来说测量另一个星系的速度不是太难：从移动光源发出的光，其频率会发生改变，如果光源远离我们，那么光就会朝光谱的红端移动（波长变长），如果光源朝向我们运动，那么光就会朝蓝端移动（波长变短）。20 世纪 20 年代的天文学家已经知道大多数星系（星云）都在远离我们运动，就是根据它们的光谱都发生红移的现象。光谱

50

的移动量依赖于光源的运动速度。通过比对观测到的星系光谱——星系发出的光的频率——与实验室中测定的已知光谱间的差别，我们很容易就能计算出星系的退行速度。

难以测量的是星系的距离。我们不能用卷尺或者激光测距仪来测量另一个星系的距离。这里，可以用三角视差法来测量最近恒星的距离，依巴谷和盖亚卫星已经将这种测距方法应用于银河系内数以十亿计的恒星，但要测量河外星系的距离，还需要天文学家伟大的天才和辛勤的汗水。近代天文学的主要推动力之一，就是建立宇宙的距离尺度阶梯，虽然有精准现代仪器的助力，关于天文学距离的争议从未平息。只要在距离测量上存在不确定因素，那么这些不确定性就会影响几乎所有其他的天文学参量——特别是哈勃常数 H。

51

由于距离测量的不确定性，哈勃自己当年给出的 H 值是现代数值的大约 7 倍。让我们再回头看看图 5，很难讲左图上那条直线就是对图中数据的最好拟合结果；右图给出了其他可能的斜率。既然如此，为什么一定要画出左图那条直线呢？

✷

你可以在厨房里更深刻地体会和理解哈勃定律到底说明了什么。找一条宽橡皮筋，用笔在上面等间距地标出一些点，代表星系 A，B，C，D，…拉伸橡皮筋，让所有标记点之间的距离都变大：A…B…C…D。

假想你位于星系 A 处。如果橡皮筋均匀拉伸，星系 B 远离星系 A 的距离是 1 厘米，星系 C 远离星系 B 也是 1 厘米，这样星系 C 远离星系 A 的距离就是 2 厘米。因为所有这些变化都是在你拉伸橡皮筋的时段里发生的，所以星系 C 远离星系 A 的速度就必然是其远离星系 B 的 2 倍。

这就是哈勃定律。

52 问题的关键是橡皮筋必须被“均匀”地拉伸，也就是每一处的拉伸率都相同。任何均匀膨胀的宇宙都会展现出哈勃定律。

我之前提到常数 H 表示宇宙膨胀的速度。准确地说，H 是宇宙膨胀速度的比值。换句话说，H 表示的是单位时间内，任意星系间距离增加的百分比。

举例来说，如果 C 最初距离 A 是 5 厘米，它每秒钟移动 1 厘米，那么 AC 的距离在 1 秒内改变了 1/5，因此 H 的数值就是每秒 1/5。橡皮筋演示可以看得更清楚，但我把公式写在脚注里。[①]

在橡皮筋宇宙中，最重要的是没有哪一个特别的星系比其他星系更能代表中心。如果你位于星系 C 处，那么你会发现星系 A 退行的速度是星系 B 的 2 倍。如果你想象一个表面粘贴着

53 星系的气球，会更清楚地理解整个画面。当你把气球吹大时，每一个星系都在远离其他星系，所有星系远离其近邻星系的速度都一样。这就是宇宙学家谈及宇宙膨胀时的精确含义。

---

① 假设星系 A 和星系 C 间的距离为 d，在星系 A 处测得星系 C 的退行速度为 v。由于橡皮筋的拉伸遵循哈勃定律即 $H=v/d$，根据定义，速度 v 是单位时间的距离改变量，通常写为 $\Delta d/\Delta t$。因此 $H=(\Delta d/d)/\Delta t$。也就是单位时间内距离改变的比值。

这样我们就有了对讲座后第一个问题的答案。我们位于宇宙的中心吗？不是。

你可能会抗议，因为气球是有中心的——在气球内部。现在我们就遭遇气球类比的失败。气球是我们三维空间中的一个两维球面，球面上的蚂蚁能够抬头看到包围的空间。我们身处的宇宙拥有三个空间维度，根本不存在能够窥视的包围空间。真实的宇宙是四维时空，没有被任何东西所包围。宇宙变得越来越大，星系之间的距离也越来越远，但宇宙并没有膨胀“进入”任何东西。这就是对讲座之后第二个问题的回答。

当然，以上所说的这些都很难想象。为了试图想象膨胀宇宙，人们通常会在头脑中假想一个有边界的膨胀的橡皮膜。一旦我们给橡皮膜加上了边界，我们就已经假设存在“外部”，而实际上根本就没有“外部”。一旦我们有了边界，就可以定义一个中心，而这也是不存在的。更好的办法是假想一个没有边界的橡皮膜，可以无限被拉伸。橡皮膜上标识的星系，永远保持彼此远离的状态。

54

*

此刻你可能会问：那么星系自身是否在膨胀呢？你我是不是也在膨胀呢？不，你和我没有膨胀（除非这膨胀来自饮食习惯，坏笑）因为电磁力将我们的身体束缚起来。那太阳系是否在膨胀呢？通常的答案是没有；太阳的引力将太阳系束缚在一起，阻止它跟着宇宙一块膨胀。类似地，星系也依靠自身引力的束缚，不会随着宇宙一同膨胀。

在更大尺度上，事情开始变得没那么明晰了，但大概是在超星系团的尺度，大约 10 亿光年，引力作用开始敌不过宇宙膨胀，不足以将质量约束在一定范围。超星系团只有一部分能被 55 引力束缚，作为整体，超星系团随着宇宙一起膨胀。因此，超星系团是宇宙里尺度最大的结构，因为任何尺度超过超星系团的结构都无法成形；宇宙膨胀阻止了结构的聚集。

✷

现在让我们回放整篇图景。如果所有星系都在彼此远离，那么假定宇宙膨胀起始于过去的某个时点就是合情合理的（尽管这并非先验的结论）。宇宙膨胀的起始事件，就是我们所称的大爆炸，天文学家弗雷德·霍伊尔为了表达对大爆炸的嘲讽，于 1949 年创造了这个词汇。

大爆炸不是通常意义的一声巨响；即使真的有人在彼时倾听，也不会听到任何响声。把大爆炸想象为发生在已有空间中的一次寻常的爆炸也是不对的。如果没有相对于宇宙的外部，那么就不存在宇宙可以爆炸进入的空间。正如我们已经知道的，时空本身就始于大爆炸。

最后，人们总说在大爆炸时，所有的物质都集中在一点，56 那里必然是宇宙的中心。但因为宇宙根本就没有中心，所以这种想法也不对。

橡皮筋能够帮助我们想得更明白。假设橡皮筋已经被拉伸了，星系 A、星系 B、星系 C 和星系 D 已经处于彼此远离的状态。现在松手，让橡皮筋回缩，所有代表星系的点都恢复原位。那么所有星系恢复原位所需要的时间，就是自大爆炸起宇宙的年纪。哈勃定律告诉我们每个星系所经过的距离 $d=v/H$。但每个星系所经过的距离也可以用它的速度乘以经历时间得到。$d=vt$，所以 $vt=v/H$，我们就得到 $t=1/H$。

哈勃常数的倒数被称为哈勃年龄，是自宇宙大爆炸以来所经历的时间估值。

此处没有任何约束条件要求全部星系（点）位于单一位置。实际上，如果我们想象这条橡皮筋是无穷长的，上面有无穷多个表示星系的点 A，B，C，…（如果英文字母数量是无穷的话），我们就必须接受这样的想法，橡皮筋大爆炸沿着这个一维表面，发生在所有地方。

认为在大爆炸瞬间，可观测宇宙中所有的物质都集中在一点是正确的。然而可观测宇宙并不是整个宇宙。自大爆炸起，光传播经过的距离称为宇宙视界，顾名思义，我们看不到任何宇宙视界之外的东西。因此我们可以说在大爆炸瞬间，宇宙视界内的全部物质都集中在一点。

57

天文学家为了测量哈勃常数而发明的各种技术，远比测量星系距离的要复杂得多。其中一部分会出现在后面的章节里。麻烦的是，这些不同测量办法得到的结果彼此并不总是一致。眼下，仅以宇宙年龄——自大爆炸起经历的时间长度——为例，140 亿年还不够准确，137 亿年才是更精确的数字。

*

广义相对论描述整个宇宙的精妙之处就在于：一旦确定宇宙中的成分及分布，剩下的就让广义相对论来告诉你宇宙是如何演化的。

这也许是广义相对论的妙方，但并不是爱因斯坦的。之前 58 我曾说过，爱因斯坦认为宇宙是静止的——非膨胀宇宙。为了让他的方程产生一个静止的宇宙，爱因斯坦不得不从数学角度强加入一个多余项：声名狼藉的宇宙学常数。它本来就是个没有意义的常数项，一旦确立了宇宙膨胀的观念，爱因斯坦就把它称为“自己一生中最大的错误”。

现在回想把宇宙学常数加入方程的经历，看起来是很奇怪的。如果火箭烟花在外太空爆炸，那么爆炸颗粒形成的云团最开始一定会迅速膨胀，假如颗粒质量足够大，那么云团的膨胀会由于颗粒之间的引力作用逐渐变慢。云团可能最终会开始收缩，这取决于颗粒的质量。但有一件绝不会发生的事情，就是云团保持静止。

同样，把没有引入常数的广义相对论方程应用于整个宇宙，结果就是个动态的宇宙。没有强加任何常数的宇宙自然会膨胀或者收缩，其速度由宇宙的物质密度决定。这实际上就是广义相对论揭示引力效应如何决定宇宙膨胀速度的主要方式。但正如牛顿物理学没有告诉我们火箭到底要携带多少火药，或者火 59 药的成分到底是什么一样，广义相对论也没有告诉我们任何宇宙的成分配方。一旦宇宙的成分密度确定了，引力就会接手剩下的事情，明确宇宙模型演化的路径。

1922 年，俄罗斯气象学家亚历山大·弗里德曼给出了爱因斯坦方程的一个动态宇宙解。由于爱因斯坦拒绝接受演化的宇宙，因此事实上是弗里德曼的宇宙模型为大爆炸理论提供了数学基础。[①] 弗里德曼宇宙的重要特征是，它是一个尽可能简单的宇宙模型。假设宇宙中的物质是均匀分布的，膨胀也是均匀的——也就是宇宙各处的膨胀速度是相同的。

弗里德曼宇宙方程精确显示了宇宙膨胀速度，亦即哈勃"常数"，是如何依赖于宇宙成分的。天文学家测得的哈勃常数其实只是"今天"宇宙的膨胀速度，技术上讲，只有在你阅读本句话的瞬间，它才真正是个常数。总体说来，随着宇宙膨胀，其物质密度会降低，宇宙膨胀速度也会随之变慢。 60

你可能还记得在第三章中曾提及，宇宙包含的物质决定空间的几何特性。如果宇宙中物质密度超过某个"临界值"，大约每立方厘米 10^{-29} 克（也就是每立方米体积内的物质质量大约是 10 个氢原子），那么就好像在空中爆炸的大型火箭烟花那样，弗里德曼宇宙的膨胀速度会逐渐减慢为 0，最终还会变为负值——宇宙将再次塌缩。这样的宇宙一般被称为闭合宇宙，其对应的空间几何特性与气球相似。

---

① 几年后，弗里德曼宇宙学模型被乔治·勒梅特（1927 年）和霍华德·罗伯特森（1935 年）、亚瑟·沃克（1936 年）分别重新发现，因此，今天的宇宙学家通常将这个宇宙学模型称为 FLRW 宇宙（取他们姓氏的首字母，译者注）。

如果宇宙的物质密度比临界值小，那么宇宙的几何特性好似无限大的薯片（上面临近的平行线会渐行渐远），将永远膨胀下去。这样的宇宙模型被称为开放宇宙。在第三章中我们知道，真实的宇宙几乎是平直的，刚好介于开放宇宙与闭合宇宙之间。膨胀速度逐渐变慢，在无限远处变为 0，现实的宇宙几乎是匍匐朝向永恒前行。[①]

61

如果宇宙未来的膨胀速度变慢，那么过去宇宙的膨胀速度就可能更快。事实上，在大爆炸瞬间，宇宙的膨胀速度是无穷大。

62

能否确定那是不可能的?

① 此处为假定宇宙学常数为 0。如果宇宙学常数不为 0（显然我们的宇宙就是这样，将在第八章中讨论这种情况），那么宇宙可能的行为会变得更加复杂。球形的“闭合”宇宙有可能永远膨胀，而“开放”的薯片形宇宙可能再次塌缩。

第五章

宇宙学的罗塞塔石碑：宇宙背景辐射

如果说发现宇宙膨胀是现代宇宙学的基石，那么发现整个宇宙都沉浸在绝对温度为 3 开的热背景中，则是现代大爆炸理论的基石。

之前我曾说过，宇宙在膨胀的事实并不意味着宇宙一定在过去的某个确定时刻始于大爆炸。很有可能一直以来宇宙都与现在看到的相差无几——这种情况下，星系仍可以彼此远离，而新的星系则非常缓慢地产生，来填补因星系远离而形成的空洞。这样的场景就是曾经非常有名的“稳恒态宇宙”，顾名思义，这样的宇宙是永恒存在的。

63

虽说想象一个永恒存在的宇宙是困难的，但同样困难的是想象一个在 140 亿年前无中生有凭空出现的宇宙。直至 20 世纪中叶，还没有任何观测证据来区分大爆炸模型和稳恒态模型。

但 1965 年，似乎一夜之间一切都变了。前一年，贝尔实验室的两位射电天文学家阿诺·彭齐亚斯和罗伯特·威尔逊用一架探测卫星回声的极其灵敏的天线，发现了来自我们银河系的射电辐射。为了精确测量，必须将本地射电干扰降至最低，无论干扰是来自电源插头还是天线本身。让威尔逊和彭奇亚斯感到不解的是，在排除了所有能想到的干扰后，还有残留的射电信号无论如何都无法消除，为此他们甚至把天线上的鸟粪都清理干净了。这个残留的微弱信号来自整个天空的各个方向，而且强度都相同，显然不可能来自银河系本身。彭奇亚斯打电话给普林斯顿大学的宇宙学小组组长罗伯特·迪克，当时他们正准备开始对这样的射电信号进行巡天观测。挂断彭奇亚斯的电话后，迪克转身对自己的同事詹姆斯·皮布尔斯和戴维·威尔金森说，“好吧，男孩们，我们被捷足先登了”。

64

彭奇亚斯和威尔逊发现的宇宙微波背景辐射（CMBR），正是宇宙大爆炸残留下来的辐射热。之前还很稳固的稳恒态宇宙模型轰然倒塌，大爆炸理论成为标准宇宙学模型。本书的其余部分将详细讨论这个标准模型是如何演化的。

✷

宇宙微波背景辐射究竟是什么？所有热物体，也就是所有温度高于绝对零度的物体，都会以热的形式向外辐射电磁能量。不仅是火炉和计算机会辐射热量，岩石、鱼类以及你我都一样。由于历史原因，物理学家把纯热辐射称为黑体辐射，把发出黑体辐射的物体称为黑体，虽然看上去它们并不黑。

黑体辐射最基本也是最显著的特征就是温度决定一切，与物体的组成无关。物体的温度告诉我们辐射量，反之亦然。在 65 等候室里，医生的助手把遥感测温计指向你的前额时，测量的就是你辐射出的热量，正因为假定你是个黑体，所以测量你辐射出的热量，也就测定了你的体温。彭奇亚斯和威尔逊将遥感测温计用于整个宇宙，测量出宇宙的温度，现在我们知道这个温度值大约是绝对温度 2.7 开。

一般的调频广播电台广播的频率大约是 100 兆赫，对应的电磁波长为 3 米。[①] 与广播电台不同，热的物体会在全波段发出辐射，但在每个频率的辐射强度有很大差别。对于黑体来说，其在各波长处辐射的能量强度——它的光谱——由黑体的温度，而且仅由其温度决定。正因如此，黑体辐射光谱看起来都很相似，大都如图 6 所示，只不过精确形状由温度决定。如图，大 66 多数的辐射集中在某个波长附近，对于温度为 2.7 开的黑体来

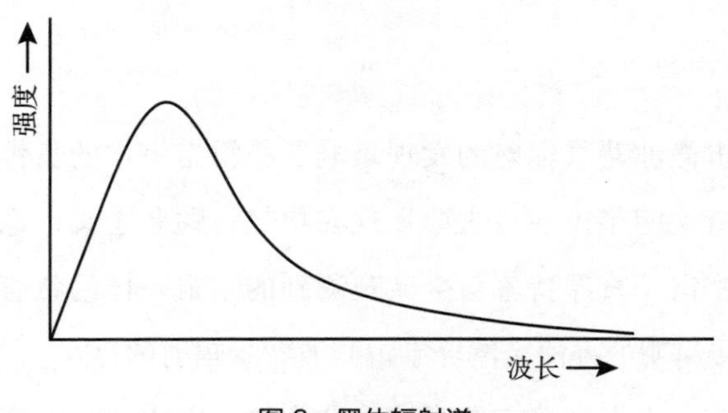

图 6　黑体辐射谱

---

① 频率和波长完全可以换算。频率 f 越高，波长 λ 越短，因为 $f\times\lambda=c$，c 就是波速（光速为每秒 3.0×10^8 米，空气中的声速约为每秒 340 米）。

说，对应的波长刚好是0.3厘米，或者对应的频率是100吉赫兹。这是电磁波谱中的微波波段，也就是宇宙微波背景辐射CMBR中字母M（Microwave，微波）的含义。

辐射强度的精确定义是每秒钟每平方厘米上流经的能量总量。好比花园水管里流出的水强一样，可以把辐射强度理解为每秒钟每平方厘米上流经的粒子数量。因为热说到底也是电磁辐射——光，这种情况下的粒子就是光子。宇宙微波背景辐射

67

的温度是2.7开，就意味着目前星系际空间中，每立方厘米体积内大约有来自大爆炸的400个光子。

自被发现以来，宇宙微波背景辐射谱已经被多个实验多次测量，最初的探测设备是1989年发射的宇宙微波背景探测器（COBE），观测结果与黑体谱的吻合度，要比人类历史有记录的任何光谱都更高。宇宙微波背景辐射就是宇宙大爆炸余晖，在我们所处的21世纪，无人怀疑。

✷

宇宙微波背景辐射的发现敲响了稳恒态宇宙的丧钟，因为它立刻就说明宇宙在过去要比现在更热。顾名思义，稳恒态宇宙认为宇宙一直保持着与今天观测到的宇宙一样的状态，显然没有直接简明的办法来解释宇宙微波背景辐射的存在。

事实上大爆炸确实非常非常非常热。因为宇宙在膨胀，其内部的物质辐射密度随着时间在降低；反过来，在宇宙的过去，

68

物质辐射密度应该更高。这其中就包括光子，在遥远的过去它

们要比今天更为紧致地被挤压在一起。

在过去，每一个光子的能量也更高。随着宇宙膨胀，从遥远区域传播过来的光波也随之被拉长，在光谱上朝红端移动。这就是著名的宇宙学红移，也经常被称为宇宙学多普勒效应，正如我在第四章中提及的。说光随着宇宙膨胀而变得更红，等同于说光子能量随着宇宙膨胀而变低。反之，在宇宙的过去，光子能量比如今更高。因为温度只是光子能量的简单测量，所以在过去，光子的温度更高。当可观测宇宙只有今天宇宙的一半大时，其温度是今天宇宙温度的 2 倍。就这么简单。

✵

这些特征带来了 3 个重要的结论。今天宇宙中普通物质的密度大约是 10^{-30} 克每立方厘米，相当于每立方米 1 个氢原子。对比来看，根据 $E=mc^2$，每立方厘米 400 个光子，每个光子的温度是 3 开，那么其质量密度大约就是 10^{-34} 克每立方厘米，只有今天物质密度的万分之一。如果不考虑其他成分，宇宙学家就可以认为目前的宇宙是物质主导的。

69

但这并非总是正确的。回望过去，物质粒子和光子的密度都以同样的速度增加，就像玻璃球挤在一个不断收缩的桶里一样。但每个光子能量都越来越高，因此当宇宙温度比今天高 1 万倍，达到 3 万开时，光子的能量密度就会超过物质密度。在此之前，大约是大爆炸之后的 5 万年时间里，宇宙是辐射主导的，意味着其行为由光子而不是物质的性质决定。这种情形可参见图 7 所示。很快，物质主导的宇宙和辐射主导的宇宙之间的差别就变得非常重要。

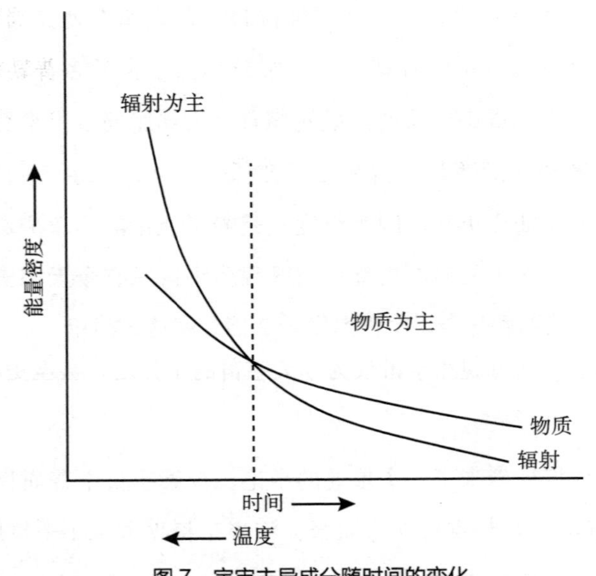

图 7　宇宙主导成分随时间的变化

不考虑扰动，一个早期炽热宇宙的第二个重要结果就是，温度会一直升高。回到大爆炸之后 1 秒时，宇宙的温度将达到 100 亿开。而在大爆炸瞬间，宇宙的温度将变成无穷高。在物理学里，无穷几乎总是个麻烦。这里的无穷高温，与第四章提到的无穷大的宇宙膨胀速度一样，都是被称为大爆炸奇点的表现形式，随着我们越接近大爆炸，奇点问题就越经常被触及。如果在时间近乎为零的时刻奇点真的存在，那就意味着理论彻底失效。这就跟算式中用 0 作除数一样——是非法的。我们得到一个无穷解，方程组就无法预言任何事情。通常宇宙学家选择大爆炸奇点稍后的时间作为思考的起点，假设那时的宇宙是可感知的，即便还不可理解。

*

早期炽热宇宙的第三个结论则是，宇宙微波背景辐射并不是精确地始于大爆炸瞬间。

可观测宇宙中大约3/4的质量都是最简单的元素氢，氢原子只有一个质子和绕其旋转的一个电子。因为电子和质子带有电量相同电性相反的电荷，所以氢原子呈电中性。

但是，当可观测宇宙仅为今天宇宙的千分之一甚至更小时，氢原子是不可能存在的。那时宇宙的温度为几千开，足以把电子“烧”得逃离原子的束缚。更准确地说，那时光子的能量足够强，能够把电子敲出氢原子，也就是“电离”它们。结果就是分离的电子和质子烩成一锅，称为等离子体。

在这样的等离子体海洋中，光子无法前行太远，因为它们几乎立即就会与电子发生碰撞，被散射掉。这就好似你在浓雾中盯着手电筒的光柱看一样：光柱被散射到各个方向，结果就是你看不到远处。在早期宇宙里，只要氢是被电离的，光子就会被有效地俘获。随着宇宙温度逐渐降低至大约3000开，等离子体足够冷时，电子才能与质子复合，形成氢原子。光子与中性原子几乎没有相互作用，在此之后——这个时期被冠以诡异的“再复合”之名，尽管跟刚开始相比，并没多复合出什么东西——大爆炸产生的光才能畅通无阻地穿越宇宙。

72

所以，我们观测到的宇宙微波背景辐射实际上是来自再复合时期，按现代测量结果，再复合发生在大爆炸之后 38 万年。在此之前，宇宙是不透明的，用任何普通的光学测量方法，我们都无法回望比宇宙微波背景辐射更早期的宇宙。记住这个术73语“再复合”。[①]

✷

第一次观测到宇宙微波背景辐射时，对于宇宙学家来说，其最重要的性质就是惹眼的均匀性。人人都一眼能看出辐射温度或者辐射强度，在各个方向上都绝对相同。此外，在足够大的尺度上，就整个宇宙而言，星系的分布也是均匀的。所有这些观测结果共同为“宇宙学原理”提供了支撑证据，宇宙学原理认为，在足够大的尺度上，宇宙是均匀且各向同性的。

有了无特征宇宙微波背景辐射谱作支持，宇宙学原理被纳入标准宇宙学模型的循环论证体系：宇宙始于大爆炸，而大爆炸就是绝对均匀且各向同性的。再没有比这更简单的宇宙图景了——但就是这么简单的模型，却有着诸多伟大的成就，下面马上就会谈到第一个成功。

这么简单的宇宙图景不可能完全正确，今天宇宙微波背景辐射最重要的特征就是它不是那么精确地均匀。1992 年，美国航空航天局发射的宇宙背景探测器（COBE）卫星在全天域的宇

① 再复合时期与通常所说的“解耦”是等效的，后者强调光子和物质之间碰撞的终止。

宙微波背景辐射谱上观测到温度极微弱的起伏（涨落），那正是宇宙学家期待的；否则，就不会有你我的存在。这些不规则的起伏代表星系形成的开始。也许你已经见过 COBE 或者其继任者普朗克卫星绘制的起伏不平的图像。传播最广的宇宙微波背景辐射图是依据 2009 年发射的普朗克卫星数据绘制的，以前所未有的分辨率展示了宇宙微波背景辐射极细微的温度变化。虽然背景辐射的温度变化只有大约十万分之一开，但这种变化的规模和分布恰恰就是解开几乎全部早期宇宙之谜的钥匙。 74

关于宇宙微波背景辐射，最重要的是什么？ 75

第六章

原初宇宙汤

碳、氮、氧、硅、铁……这些是我们日常生活中习以为常的元素，也是生命所必需的。但我们要清醒地认识到，所有这些元素加起来，还不到可观测宇宙质量的 1%。可观测宇宙的绝大部分，大约 76% 是由最轻的化学元素氢构成的，而第二轻的元素氦，则构成了其余的 24%。天文学总能正确地看待事情。

76

20 世纪最伟大的天体物理学成就之一就是意识到恒星其实是核反应堆，正在把氢变为更重的元素，包括以上提及的种种。有时通过超新星爆发，所有这些元素都会被撒入太空，进而形成更重的元素，例如铅、金还有铀。最终，这些重元素会混入婴儿时期的太阳系、行星以及我们自身。

本质上来说，我们对于恒星成分的全部理解都来自对其光谱的观测。任何光源的光谱一般都包括独立的线系，构成光源的不同化学元素会在不同频率发光。例如，虽然地球上绝大多数的氦都来自深埋在地表之下的放射性元素衰变，但在恒星光谱中也观测到氦。实际上，氦源于希腊语“赫利俄斯（太阳神）”，首先于 1868 年在太阳光谱中被发现。对最古老恒星的现代观测表明，它们的质量构成有 24% 的氦，以及痕量的其他轻元素。

由于构成最古老恒星的氦以及其他痕量轻元素在恒星形成时就已经存在了，那么问题来了：这些元素是如何形成的呢？ 77

20 世纪 40 年代后期，物理学家乔治·伽莫夫和他的同事为了精确地回答这个问题，提出了如今被称为“热大爆炸”的理论。最终，这个问题也的确得到了答案。热大爆炸理论能够精确预言宇宙中各种轻元素的丰度，使得它成为继发现宇宙膨胀和宇宙微波背景辐射后，早期大爆炸理论的第三个胜利，也是整个大爆炸理论的基石之一。

*

早期宇宙中轻元素形成的理论，被称为大爆炸核合成，或者更诗意地称为原初核合成，不仅因为它成功地解释了观测到的元素丰度，更重要的是，它代表了广义相对论与核物理学的成功融合。同时，原初核合成还对第五章末尾提出的问题给出了第一个答案，那就是为什么宇宙微波背景辐射对于宇宙学来说如此重要。事实上，即使在微波背景辐射被探测到之前，伽莫夫和他的同事的计算结果中，已经假定这种宇宙热背景必然存在。

原初元素形成的宇宙汤就是第四章中提到的弗里德曼宇宙， 78
其假定宇宙中的物质能量均匀分布，且正在膨胀，膨胀速度又由其中的物质能量所决定。以最粗线条的视角来看，整个元素形成过程非常简单：由最初的膨胀的宇宙汤开始，加入必需的成分，就可以上菜了。

此前我曾说服你相信宇宙在过去要比今天炽热得多。实际上，在大爆炸之后几分钟内，宇宙的温度足够高，核聚变反应能够进行，与太阳内部发生的反应一样，都是在把氢转变成氦。随着宇宙膨胀，温度降低，“在不足以把一个土豆烧熟的时间里”，伽莫夫这样描述，整个核聚变反应就停止了。结果就是观测到的宇宙中留下 24% 的氦以及其他痕量轻元素。

这个听起来高大上的概念，既不精确也不完整，所以让我们花点时间在细节上，那才是麻烦所在。最重要的事是要牢记这里没有任何猜想，全部图景所需的仅仅是常规物理学。

老实说，技术上讲，我一直在说的是同位素。元素是根据其原子核内的质子数目来区分彼此的，而对于某种元素来说，由于其原子核内的中子数目不同，就有了同位素的概念。普通的氢原子核只包含 1 个质子，而氘（“重氢”）的原子核则包含 1 个质子和 1 个中子，氘就被称为氢的同位素。普通的氦原子核由 2 个质子和 2 个中子构成，被称为氦 –4，而氦 –3 则是氦 –4 的同位素，因为它的原子核由 2 个质子和 1 个中子构成。

79

我们的目标是要在一个非常热的火炉中，产生与天文学观测相同丰度的这些同位素。首先，我们需要原材料。为了让食谱配方简单，我们假设早期宇宙中的物质就是构成今天各种化学元素的最基本成分：中子、质子和电子。烹饪过程完成的环境则是每立方厘米 400 个光子（见第五章），也就是宇宙微波背景辐射。

还需要一种原料：被称为中微子的亚原子粒子。中微子是所有基本粒子中，除光子以外质量最小的粒子，在自然界，它们几乎不与其他粒子发生相互作用。一个中微子能够在被捕获前，畅通无阻地穿过厚达 1 光年的铅。因此，那些大爆炸产生的中微子，至今还没有被直接探测到。然而，我们知道它们必 80 然存在的原因之一就是，如果没有这些中微子，那么整个核合成进程根本不会发生，更不用说给出正确的答案了。

✳

这就是全部的原料清单。接下来，火炉的温度必须合适。为了避开大爆炸奇点，那时的温度无限高，我们要选择一个非 0 的时间作为起点。就从大爆炸后 0.0001 秒开始吧。从今天宇宙微波背景辐射的 2.7 开回溯，那么大爆炸后 0.0001 秒时，宇宙的温度大约为 1 万亿开。

谈论这些数字似乎是异想天开，在物理学中，万分之一秒内能发生很多事，而 1 万亿开的温度虽然很高，但也并非不可想象。质子和中子能够在 1 万亿开的温度下存在，它们彼此之间的核反应对于物理学家来说再普通不过。如果温度更高，那么中子和质子就会“蒸发”成为构成它们的夸克，任何核反应都不再存在。

1 万亿开对于原子核来说还是太热了。质子和中子在这锅粒 81

子汤里飞速运动，第一章中介绍过的强核力无法把它们束缚在一起形成氘核或者是氦核。正如在几千开时原子氢会被电离形成质子和电子构成的等离子体，当温度为 1 万亿开时，原子核也会被“电离”成由中子和质子构成的等离子体。

但是，大约 1 秒后，温度降到只有 100 亿开，大约与太阳中心的温度相当，这时冷却已经足够，原子核能够开始形成了。假设在大爆炸 1 秒后，宇宙中每个中子（n）都对应存在 7 个质子（p），如图 8 所示。

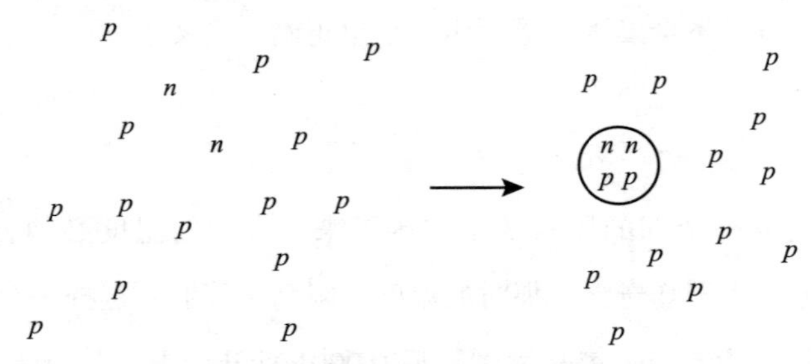

图 8　大爆炸后 1 秒中子与质子的比例

82　在大爆炸之后近乎 3 分钟时，宇宙温度继续降至 10 亿开，这时发生碰撞的中子和质子就能够粘在一处，形成氘核（np）。随后的一系列核聚变反应，实际上与太阳内部或者地球实验室里发生的一模一样，氘核迅速转变为氦 -4，也就是普通的氦（ppnn）。[①] 氦是一种异常稳定的元素，核聚变反应也在此处终结。

---

① 主要的核反应是：$n+p\rightarrow d$；$d+d\rightarrow {}^{3}He+n$；$d+d\rightarrow t+p$；$t+d\rightarrow {}^{4}He+n$；${}^{3}He+d\rightarrow {}^{4}He+p$；$d+d\rightarrow {}^{4}He$。其中 d 代表氘核；t 代表氚核（“超重氢”），包含 1 个质子和 2 个中子。

所有这些都发生在大约 1000 秒内，然后一切都稳定下来——也许还不足以煮熟一个土豆，当然这取决于土豆的大小。

那么到底生成了多少氦呢？如果在大爆炸之后 3 分钟时，每个中子对应有 7 个质子，而所有的中子都转变为氦，那么核反应就会在中子耗尽时停止。从图 8 中可以看出，结果是每 12 个氢核（质子）能够产生一个氦核。但由于氦核的质量是质子的 4 倍，这就意味着从质量上说，我们得到 75% 的氢和 25% 的氦，这个结果与真实宇宙的观测值相近。

当使用计算机进行精确计算时，结果表明痕量的氘以及其他同位素也同时产生，图 9 给出随着宇宙温度降低，各种轻同位素的质量占比是如何演化的。如果说仅仅正确预测氦丰度就

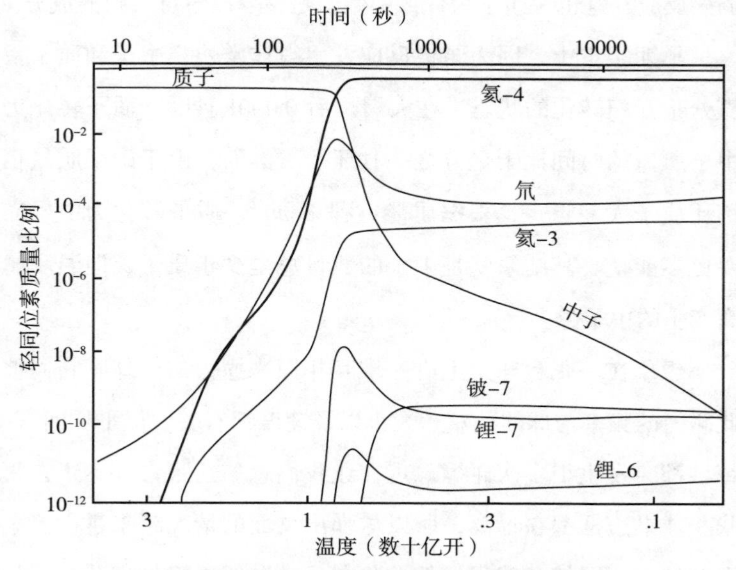

图 9　随宇宙温度降低，轻同位素质量占比变化

是一项重要成就的话，那么所有轻同位素丰度都与天文学观测值吻合的结果，就太令人惊艳了。这项近乎奇迹的发现正是宇宙学家坚信大爆炸理论的主要原因之一。

84

*

此刻，我希望你能提出这个问题：为什么中子和质子的比例刚好是 1∶7 呢？要说明这一点并不十分困难。

首先，中子和质子是可以互相转化的。1 个中子本质上就是 1 个质子加 1 个电子：$p+e \rightarrow n+\nu$，其中 ν 代表中微子。反应方程也可以逆向进行，中子可以转化为质子 $n+\nu \rightarrow p+e$。由于这些反应都由第一章中提到的弱核力决定，因此它们被称为弱核反应，这也说明了为什么中微子是核合成不可或缺的成分。

早期宇宙中，因为弱核反应发生得非常快，中子和质子一直处于互相转化的状态。在大爆炸后 0.0001 秒时，质子转化为中子所需的时间比十亿分之一秒还短。然而，由于中子质量稍稍比质子大一点点，根据质能方程 $E=mc^2$，质子转化为中子需要更多能量。其结果就是中子的数目总是少于质子，但温度越高产生的中子越多。

85

想象在一张台球桌上有一堆互相撞来撞去的台球。它们彼此碰撞的频率与球的数量、大小以及速度都有关，但平均而言，每秒都会发生很多次碰撞。现在让我们想象这张台球桌正在膨胀。球桌边框都在远离，所以球弹出桌面的情况越来越少。桌面被拉伸，即使台球仍然彼此相向运动，但碰撞越来越少。如果桌子膨胀得足够快，那么所有的碰撞都会终结。

无论是在生活中还是在物理学范畴，当两种尺度交叉时，就会不可避免地发生有趣的事。由公众基金支持的大型项目可能耗时几十年才能完成，但美国联邦政府每 4 年就要换届；尺度交叉，项目取消，混乱难以避免。

早期的宇宙很像一张不断膨胀的台球桌，其膨胀率完全取决于其中所包含的成分。从今天宇宙的测量值回溯，大爆炸刚刚发生不久时，宇宙的密度完全由光子和中微子主导。相比而言，中子和质子的密度太小，因此对宇宙膨胀速度的贡献可以 86 忽略不计。用第五章的话来说，此时的宇宙完全是辐射主导的宇宙。

在大爆炸之后 0.0001 秒时，由弱核反应主导的中子 - 质子转化速度，要比宇宙膨胀快大约 100 万倍。只要弱核反应得以持续，宇宙就有可能根本不会一直膨胀。

但情况很快就发生了变化。随着温度降低，弱核反应速度急剧变慢，大爆炸后大约 1 秒时，反应速度就慢于宇宙膨胀的速度了。中微子停止了与中子和质子的碰撞，就好像台球桌上那样，弱核反应也停止了。此刻宇宙里中子与质子数目的比例刚好是 1 : 7，这个比值也因为反应停止而被“冻结”住，在之后大约 3 分钟开始的核合成之前，不会发生太多变化。[①] 剩余的 87 中子继续核合成，如前文所述，直到中子耗尽，留下 24% 的氦。

---

① 自由中子是放射性粒子，会发生衰变，其半衰期大约是 10 分钟。因此，在核合成开始时，大约 20% 的中子已经衰变。中子衰变过程见图 9。

不要忘了目前为止所有这些讨论仅考虑了原子核。原子则直到 38 万年后的再复合时期才会出现，那时宇宙的温度下降得足够低，电子得以与原子核结合形成原子。

宇宙中最终的氦丰度几乎完全由“冻结”时中子与质子的数目之比决定，20 世纪 80 年代的宇宙学家根据这一点，在建立专门的实验室之前就预言了自然界中存在的中微子种类和数量。虽然我们已经知道有 3 种中微子，它们被称为中微子的“味”，但并不排除还有更多味存在的可能。任何一种额外中微子味的存在，都会大大提高核合成时期宇宙的膨胀速度，进而提高氦丰度（因为宇宙膨胀速度会在更早期更高温度时就超过弱核反应速度，那时的中子数目也更多）。

88

所以增加中微子种类就意味着更高的氦丰度。将氦丰度限定在观测到的 24%，就排除了新的中微子类型，这个预言在后来建造的地面粒子碰撞实验室里得到证实。

✷

除了确实能解释很多观测结果，也许原初核合成理论最不寻常的一点就是没有任何含糊其词的地方。大爆炸后 0.0001 秒的初始条件都是普通物理学的范畴，所涉及的核反应也都是实验室中熟知的。在全部场景里，只有一个数字是可调的：今天宇宙里中子和质子的密度，这个数值在核合成发生时就已经固定下来。因为中子和质子总称为重子，所以宇宙学家把它称为今天宇宙的“重子密度”。

如今，只说某种疾病的致死数量并不如告知它的致死率传达的信息更多。对核合成理论而言，这个单一的输入值可以被表示为光子与重子的数目之比。我们宇宙中光子与重子数目的比值大约是 $10^9:1$，也就是每个重子对应 10 亿个光子，正是这个数值让核合成理论得到与观测完全吻合的结果。然而我们并不知道为什么光子与重子数值比刚好是 $10^9:1$ 而不是 1 或者 618。可能宇宙只是恰好就以这样的光子 - 重子比开端的吧。物理学家都是怀疑论者，认为这个事实是精细调制的结果——换句话说，就是精心调制模型参数使之与事实相符。他们更希望找到一种自然的机制来解释光子 - 重子数之比的起源。

“很自然”地，我们会期望宇宙初始时物质和反物质数量相等——没有根本原因认为一种要比另一种更好——但我们的宇宙几乎完全是由所谓的“物质”构成的。① 1967 年，物理学家安德烈·萨卡洛夫提出，在大爆炸时，物质和反物质数目产生了极细微的差异——每 10 亿个反物质粒子对应 10 亿零 1 个物质粒子。《星际迷航》的粉丝们一定都知道，当物质和反物质接触时会发生湮灭，每次湮灭产生两个光子。如果每 10 亿个物质粒子都与 10 亿个反物质粒子发生了湮灭，那么就会剩下 1 个物质粒子。我们就生活中这“剩下”的宇宙中，每个重子周围都环绕着几十亿个光子。这个解释只是把问题向后推了一步：又是什么决定了物质与反物质数目微弱的不平衡呢？

① 例如反质子和正电子，它们分别拥有与质子和电子相同的质量，只是所带电荷相反。

虽然萨卡洛夫提出了物质反物质不平衡的必要条件，但仍然无法解释观测到的光子重子数比值。这成为物理学的未解难题。

总的说来，我们还搞不清楚物理定律是怎么形成的。天体物理学的最大成功就是证明我们有关动量能量守恒的假设，目前为止在宇宙各处仍然有效——宇宙学中类似原初核合成理论的成功则在于，自大爆炸以来，自然规律还没有发生重要的改变。

数学家艾米·纳托提出的基本原理告诉我们，如果一个系 91 统不随时间发生变化，那么它的能量保持恒定，即守恒；如果空间是均匀的，那么这个系统的动量（质量与速度的乘积）也守恒。但这个原理并不能解释空间是如何变得均匀的，而且还引发了新问题，我们是否能够用日常的物理学定律（见第十一章）建立极早期宇宙模型，那时宇宙还没达到均匀的状态。而且当我们说“能量既不能产生也不能消失”时，我们考虑的总是面包盒子那样的闭合有限的系统。面包可以转化为能量，代价是面包质量减小；但对于整个宇宙来说，特别是如果宇宙是无限的，讨论能量守恒又意味着什么并不是很清楚，如果有的话。

92

我们能否避免精心调制宇宙?

第七章

暗黑宇宙

讲座之后，公众很少提及有关原初核合成的问题。通常取而代之的是：“你能告诉我暗物质是什么吗？”

答案应该很直接：不能。

本章结束。

实际上当然不能这样，所以让我们重新思考。按照爱因斯坦从未说过的“名言”“让事情尽可能简单，但又不能太简单”，物理学家的工作就是砍掉自然界的枝杈，找到对所观测到现象的最简单解释。然而自然鲜少会像第一眼看上去那样简单。随着观测到的现象越来越复杂，用来解释现象所必需的模型和理论也会从简单演化为复杂。然而，与经济学家不同的是，物理学家只是情非得已时才为理论和模型添加复杂因素。

93

1965 年以后，大爆炸理论已经被广泛接受，假定物质分布绝对均匀的弗里德曼宇宙成为标准宇宙学模型。但 COBE 卫星发现在宇宙微波背景辐射中存在些微的涟漪，迫使标准宇宙学模型必须修正，从而能够解释星系、星系团以及超星系团无可否认的存在。

后面的第九章和第十章中，我们会遇到新的标准模型，但在此之前，我们必须先承认暗物质和暗能量的存在，它们是新标准宇宙学模型的部分基础。暗物质和暗能量都是危机四伏的话题，巨大变化的时标是以周来计数的。这种情况下，采用《纽约时报》的原则是非常明智的：如果《纽约时报》声称有了某项发现，而此前你并不曾从该研究领域专家那里听说过，那么就不要相信。

*

通信卫星之所以沿着环绕地球的圆轨道运行，只是因为引力将它们的运动轨迹束缚成闭合轨道，引力抵消了卫星的自然惯性，因为按照惯性定律，卫星会沿直线飞出地球进入深空。由于卫星受到的引力取决于地球的质量，因此卫星的轨道速度也由地球质量决定。卫星的轨道速度越大，将其维持在轨道上所需要的质量也越大。绕太阳运动的行星或者绕银河系中心运动的恒星，都遵循同样的理论。

94

在过去的一个半世纪里，存在看不见物质的设想几次涌现。20 世纪 30 年代，天文学家弗里茨·茨威基注意到星系团的成员星系轨道速度太大，无法用星系团内的发光质量——恒星——来解释，因此他提出一定存在所谓的暗物质，来补足缺失的星系团质量。当时，暗物质被简单地定义为不发光物质。直到 40 年后，茨威基的想法才被认真对待，因为此时维拉·鲁宾发现星系边缘处的恒星轨道速度也过大，无法用星系内的发光物质质量来解释。如果星系质量仅仅是全部发光物质的总和，那么边缘恒星就应该飞入星系际空间，而不会绕星系中心运动了。

95

鲁宾和她的团队采用了简单直接的办法进行测量。根据多普勒位移，很容易就能测定绕星系中心运动的恒星速度。截至目前，已经对成千上万的星系和星系团进行了这样的测量，观测结果十分一致：星系中大多数物质都是不可见的。实际上，宇宙中大约 85% 的物质看起来都是暗黑无光的。

这样的观测结果牢不可破，而讲座之后的问题也非常简单：暗物质由什么构成？答案也同样简单：我们不知道。任何给出与此不同答案的人，要么是推销员，要么是政客，但绝不是科学家。

任何不发光的物质都曾被认为是暗物质候选体。这样的物质实在是太多了，这本小书不可能全部讨论——事实上任何一种都没法讨论，因为尚未发现任何没有被排除的暗物质候选体。

✷

有两个天然的暗物质候选体，一个就是黑洞，因为按照定义，黑洞不发光，另一个是黑洞的表亲，中子星。还有可能是“褐矮星”，它们是“失败的恒星”，质量大约为木星的几

96

十倍。褐矮星只能发出非常微弱的光，因为它们的质量不足以启动核聚变反应。暗物质也可能是木星——很多很多木星——它们也贡献一部分暗物质质量。天文学家把这样的天体统称为

MACHOs——晕族大质量致密晕天体。不幸的是，MACHOs 基本上已经被排除在暗物质候选体范围之外，当然是有足够理由的。

正如在第三章所说，按照广义相对论，光经过大质量物体时会发生偏转。因此光经过恒星、黑洞或者星系周围时，将偏离原来的传播路径，这与光经过普通透镜会发生偏折是完全一样的。这种“引力透镜”造成的结果就是，位于大质量透镜天体后方的天体，其光学像会偏离本来的位置，或者被扭曲变形。今天，人们已经确信引力透镜的存在，包括哈勃空间望远镜在内的现代望远镜拍到了众多姿态万千的壮观景象。

由于银河系在自转，位于星系边缘的晕族大质量致密天体也随之旋转。如果银河系外（河外）光源，例如一颗非常明亮的恒星，发出的光经过某个作为引力透镜的致密天体，那么当透镜天体运动到明亮恒星前方时，我们就会看到那颗恒星光轻微闪烁。对银河系以及大小麦哲伦云进行的引力透镜统计研究表明，没有任何有力的证据支持晕族大质量致密天体的存在。

97

将晕族大质量致密天体排除的更确定的原因则是原初核合成。无论 MACHOs 到底是什么，它们都是由普通重子物质（中子和质子）构成的，这些物质在核合成时期就已经存在了。增加重子密度就会提高核合成时期形成氦元素的核反应速度，也就会生成更多的氦。而天文学家观测到的氦丰度显示，重子密度与宇宙中的发光物质是相对应的。如果宇宙中真有 5 ~ 6 倍的暗物质，那么它们绝不可能来自重子物质，因为那样的话，大爆炸就会产生太多的氦。科学理论总会从多个方面互相印证，这就是一个最好的典型实例。

另外，在第十章中我们还会看到，与核合成理论一样，COBE 卫星探测到的宇宙微波背景辐射的涟漪，对暗物质与重子的比值提出了具体的约束。不管暗物质究竟是什么，反正不是构成你我的普通物质。

98

✷

鉴于上述原因，很自然地就会将中微子作为暗物质的下一个候选体。光子，是电磁力的传递者。中微子则是在弱核力环境中产生的，它们不是光子。然而它们确实很轻。事实上，半个多世纪以来，物理学家一直认为中微子与光子一样，完全没有质量，所以当然也就没有考虑中微子是暗物质候选体的可能性。

但是，自 1998 年起观念开始发生变化。日本的超级神冈中微子探测器发现，在第六章里提到的三种“味”的中微子，可以通过振荡实现相互转变。这种振荡就好像当你敲击钢琴琴键，而钢琴稍稍有点走调时听到的声音。不同频率的声音来自不同的琴键，同样，中微子不同的振荡速度源于不同“味”的中微子拥有不同的质量。如果中微子质量为 0，那么也就不存在中微子振荡了。

正是因为中微子振荡确实存在，所以我们知道中微子是有质量的。不幸的是，中微子实在是太“羞涩”了，为了测量到

99 它们的精确质量，实验物理学家挠头努力了几十年。中微子振荡实验给出了极细微的质量差异，也意味着中微子质量极小，为了探测中微子质量而设计的物理实验，其实更直接地显示出中微子质量至多是电子质量的五十万分之一，如果不是有中微子，电子就是已知质量最小的粒子了。这进而说明中微子质量的最大值，也只是中子或质子质量的二十亿分之一。而普朗克卫星对于宇宙微波背景辐射涟漪的观测结果显示，中微子的质量要比这个值还小。

因此，即使在最乐观的情况下，中微子质量也小得令人难以置信。但要记住：每个重子对应着大约 10 亿个光子。因为中微子与重子的数量比值几乎跟光子 - 重子比相当（实际上是稍微少一点，具体的数值取决于中微子质量），因此中微子的总质量能达到重子总质量的几分之一。我们所处的 21 世纪 20 年代，任何事都难有绝对，但似乎不大可能指望中微子贡献超过百分之几的暗物质。

100 物理学中总是有很多“但是”。不排除有第四种中微子存在的可能，它不与其他三种中微子发生振荡，却有着更大的质量。这种中微子被称为“惰性中微子”。但由于目前还没有关于惰性中微子确切的证据，我就不骚扰它们了。

*

几十年来，呼声最高的暗物质候选体其实一直都是弱相互作用大质量粒子（WIMPs），而不是晕族大质量致密天体

（MACHOs）。与中微子一样，WIMPs 也不经由电磁力与物质相互作用——也就是说，它们既不发射也不吸收光——因此它们可能真的是暗物质。一般认为它们有质量，介于 10 倍到 1000 倍中子或质子质量之间，因此能够与普通物质通过引力相互作用，或者可以直接碰撞。WIMPs 理念的弱点是，它们完全是假想出来的。

对 WIMPs 的搜索已经持续超过 20 年。一个典型的 WIMPs 探测器就是充满低温氩气或氙气的冷冻箱。一个 WIMPs 能够与一个氙原子发生碰撞，产生微弱的闪光，可以被冷冻箱中的探测器探测到。有两个主要的困难：首先，WIMPs 不是唯一能产生碰撞闪光的粒子，宇宙线或者来自附近放射性元素衰变的粒子都能够产生同样的闪光，这种“假信号”必须被排除。因此 WIMPs 探测器总是被深埋在地底，通常是废弃的矿井，就是为了屏蔽可能会造成干扰的背景。第二个困难是没有人确切知道到底要搜寻什么，对于实验设计者来说，如何标记“嫌疑犯”就成了最大的挑战。

101

直到目前，WIMPs 的搜寻均告无果。2020 年当意大利的 XENON1T 探测团队认为他们可能探测到了一个轴子时，兴奋的心情飘忽而过。

轴子被很多人视为最后、最有希望的暗物质候选体。轴子的英文名 Axion 是一款洗洁精品牌，这种粒子是 20 世纪 70 年代粒子物理学家为了解释强核力的难题（具体来说，构成中子的夸克都是带电荷的，但为什么中子却表现为电中性呢？）而构想出来的。轴子非常轻，甚至比中微子还轻，但在某些早期宇宙的场景中，会产生足够数量的轴子，它们的总质量能够满足暗物质质量的需求。然而，这样的早期宇宙场景本身都还存疑，而且任何有关轴子的事好像都以错误告终，所以我们还是把轴子放在一边吧。

102

✷

如此之多的负面结果，如此之多的假设猜想，如果科学家不曾想过用其他理论来代替暗物质，那才叫怪事呢。确实有一些宇宙学家反对暗物质概念，认为应该用修改后的牛顿引力理论替代它。在星系边缘，引力似乎太弱，无法将恒星束缚在轨道上。而实际上，牛顿的引力定律并没有在这么大的尺度上被检验过，因此为什么不能只是简单地修改一下定律，让外围的引力变得强一些呢？这种策略被称为修改的牛顿动力学（MOND）。

确实可以重写牛顿引力定律，来解释星系边缘恒星的轨道运动问题，但这就需要引入一个特别的长度，超出这个长度以外的区域，引力要比牛顿定律描述的更强，而这个长度等同于引入了一个新的自然常数，就像光速或者电子质量那样。物理学家是最不愿意做这种事的。而且因为牛顿定律就是广义相对论在日常生活中的边际，任何修改的牛顿动力学都意味着要修改广义相对论。物理学家做了很多这方面的努力，但目前为止看起来所有的努力都与观测结果不符。可以说，在绝大多数宇

103

宙学家看来，修改的牛顿动力学远比暗物质的想法更不可信。

✷

读完本章你可能会觉得与其说是有关宇宙学的，不如说是有关粒子的。某种意义上讲，确实如此。宇宙就是个极端现象频现的大舞台，至少在我们的时代，无法将宇宙学和基本粒子物理割裂开来。广义相对论、核物理、基本粒子物理以及更多物理学分支都交织在一起，共同组成当前我们的宇宙图景，缺少哪一条丝线，都无法织就这幅宇宙全图。我们要搞清楚，物理学中任何新提出的想法，必须与 400 年的实验和观测结果相符，而自然总是比我们更聪明，一直如此。

你是不是忘记暗能量了？　104

第八章

更暗黑的宇宙

不是，我没有忘记暗能量。

意识到构成你我的物质只占宇宙物质总量很小的一部分，着实让人头脑清醒，更进一步，意识到整个宇宙甚至可能根本不是由物质构成的，就让人更加冷静。过去的 20 年里，主流天文学家和宇宙学家都已经接受了这样的事实，当前宇宙的绝大部分并非物质，而是暗能量。“暗能量”这个词不比一个占位符更有意义；除了知道它不是物质和大约占据 70% 的宇宙能量，我们对其一无所知。

105

也许本章也应该在此画上句号，但为了理解为什么大多数宇宙学家相信暗物质存在，我们就必须相信第四章中提到的哈勃定律是个谎言。根据哈勃定律，遥远星系的速度 - 距离关系可以用一条直线表示，而这个关系成立的前提条件是宇宙要一直保持匀速膨胀才行。那种情况下，哈勃常数 H 才真正是个常数，哈勃定律 $v=Hd$ 才能成立。

另一方面，我们想当然地会认为星系彼此间的引力吸引必然会让宇宙膨胀慢下来。因此，最遥远星系（它们发出的光从宇宙早期就开始传播，现在才被我们接收到）的退行速度应该比哈勃定律描述的更快。这种结果如图 10 中标明减速膨胀宇宙的线条所示。

1998 年，两个研究团队（超新星宇宙学项目和高红移超新星研究团组）独立宣布宇宙并不是在减速而是在加速膨胀，轰动一时，全世界的宇宙学家都震惊了。宇宙过去显然在加速。宇宙学家曾打赌，这个结果会跟大多数不可信的物理学结果一样灰飞烟灭，但来自彼此竞争的研究团队的证据让结果变得十分可信，要知道竞争者是宁可让对方毁灭也不愿彼此合作的。至今，它们经受住了时间的检验。 106

从概念上讲，研究团队的工作其实非常简单。跟哈勃一样，他们也把很多星系的速度和距离画在一张图上，试图寻找数据点相对一条直线的偏离。如图 10 所示，这种偏离不可能在近距离处找到，因此研究人员需要测量遥远星系的距离，其距离之

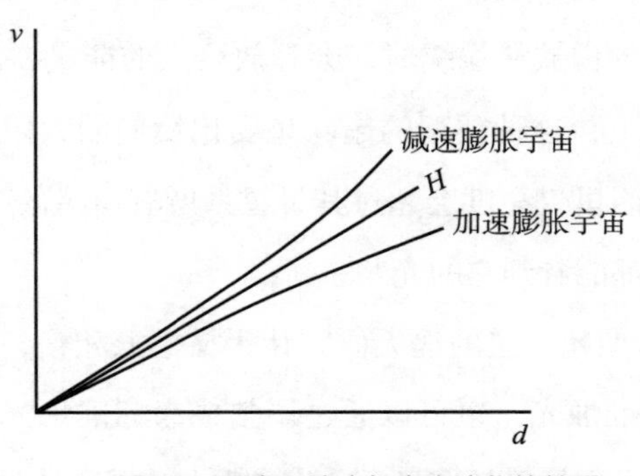

图 10　距离、速度与宇宙演化的关系

远可与可观测宇宙相比。

解决众所周知的测距难题的关键，是要找到一种可信赖的 107 标准烛光。日常生活告诉我们，同一个灯泡距离越远，看起来就越暗。具体说来，灯泡的视亮度与其距离的平方成反比：如果距离加倍，那么亮度就减为 1/4；如果距离增加 4 倍，那么亮度就减为 1/16，以此类推。

假如我们看到两个灯泡，其中一个比另一个暗 4 倍，我们就面临两难的选择：有可能我们看到的是并肩摆放的两个灯泡，一个 25 瓦和一个 100 瓦；但也有可能，两个灯泡都是 100 瓦，只不过其中一个的距离是另一个的 2 倍。但如果我们碰巧知道两个灯泡是同样亮度，那么就可以肯定其中之一距离是另一个的 2 倍。而且如果我们能确知每个灯泡都是 100 瓦，我们就能准确知道灯泡发出的能量是多少。反过来，如果我们能测量出接收到的能量是多少——视亮度——我们就能推算出灯泡的距离。

所谓标准烛光，简单说来就是一个我们知道具体瓦数的灯泡。超新星项目选择的标准烛光就是 Ia 型超新星。当一颗从其 108 伴星吸积物质的白矮星塌缩时，会释放巨大的能量，成为 Ia 型超新星。实际上，这种类型的超新星要比我们的太阳亮几十亿倍，在几天时间里，亮度会超过其母星系所有恒星的亮度总和，在整个宇宙中都能看到它的光芒。

对众多 Ia 型超新星的巡天研究让天文学家相信，即使它们不是精准的标准烛光，也可以通过调整成为标准烛光。利用 Ia 型超新星数据得到的哈勃图显示，宇宙看起来是在加速膨胀。

✷

加速膨胀意味着一定存在某种力量，正在将星系彼此推开。通常这种力被称为“反引力”，但没有任何意义。无论这种力是什么，它并不是引力的反作用力。曾有一段时间，宇宙中这种神秘成分被称为“精质”——亚里士多德的第五元素——只是掩饰无知的优雅词汇。最近，它又被冠以“暗能量”的标签，对其本质的解释并没有任何增益，不要与前一章所讲的暗物质混淆。暗物质与暗能量没有任何可观测到的联系。一种是物质，109 而另一种嘛，是能量。

与暗能量最近似的是第四章提到的爱因斯坦的宇宙学常数，是他为了保持宇宙稳恒，人为加在场方程中的因子。因为爱因斯坦当时使用了希腊字母拉姆达 Λ 来表示这个因子，今天的宇宙学家通常就把暗能量称为场方程中的“拉姆达”。与引力不同，宇宙学常数的确是个常数，不随宇宙膨胀而改变。与稳恒态宇宙学相反，在我们的宇宙中，暗能量 Λ 施加一个向外的压力，使得宇宙加速膨胀。

我们并不知道宇宙学常数从何而来。被广泛接受的猜想是其代表了时空的真空能，是大爆炸自身留下的产物。根据量子力学，真空并不是真的空无一物，而是可以被视为喧闹的能量海洋。在物理学家的头脑中，这个能量海洋描绘的是由代表光子、中微子和其他粒子的不断振荡的极细小的弦所构成的场。你可能听说过著名的海森堡测不准原理，那是自然的一条法则。110

它告诉我们，不可能同时精确测定一个粒子或一条弦的位置和速度。一条弦的能量取决于它的延展（位置）和振荡速度。根据海森堡原理，这两个物理量不可能同时为 0，因此真空里的弦总会拥有一些能量。

困难在于，如果我们估算宇宙起始时总的零点振荡能量，会发现其数值要比今天的暗能量至少大 120 个量级。由于能量不会变化，因此直到今天，它一直保持比暗能量大 120 个量级。这就是所谓的宇宙学常数难题。

宇宙学家就面临两个选择：要么暗能量 Λ 不是量子涨落的结果，这种情况下，无人知道它如何而来；要么必须找到一种办法，能让真空能减弱到今天的观测值，大约是物质密度的 15 倍。当然，如果 Λ 的值要比今天大 120 个量级，我们所知的宇宙就不可能存在。如果那样的话，宇宙会膨胀得太快，以致星系根本无法形成，原初核合成也从未发生。

111

因此，如果你相信宇宙学常数的初始值像以上简单估算的那么大，那么就必须发现一种机制能让它显著减小，而且要非常迅速。各种努力尝试仍在继续，但至今为止还没有令人信服的结论。

总是存在第三个选择。最近，一些宇宙学家质疑 Ia 型超新星作为标准烛光测距的资格，这就意味着观测到的宇宙加速膨胀的结果可能是错的，那么就不存在暗能量。这真是对这个难题的优雅解决之路（堪比 2011 年因宣称发现超光速中微子而引起的骚动。后来证明不过是虚惊一场，只是因为探测设备某个连接处松动了）。虽然还有些宇宙学家另有质疑暗能量的理由，但目前这些声音仍是少数。我本人不会加入混战，我还希望在墨水没干之前，能让本书尽享天年呢。

其实至少还有另一个选择。如果宇宙学常数太大以致星系都不能形成，那么几乎可以肯定，宇宙中也就不会有生命。而我们在这里提出问题这个明显的事实，就理所当然地要求宇宙学常数必须减小。这就是“人择原理”的一个示例，我们将在第十五章中继续讨论。 112

✷

你可能已经注意到，宇宙学常数问题与第六章提出的光子-重子比为什么是 10 亿比 1 的问题非常相似。两个问题都需要对数值的起因做出解释，而又没有明显的理由说明为何要有特定取值。你也许感觉到，这种难题本质上与试图确定哈勃常数大小的难题是不同的，后者纯粹是个观测问题。

确实如此。重子-光子比和宇宙学常数问题更像是“为什么”难题，而不是“怎么样”难题。传统上讲，科学的范畴是“怎么样”而不是“为什么”，但在过去的一个世纪里，情形发生了变化，观测与理论之间的界线越来越宽，理论物理逐渐变为思考“为什么”的科学。

这样的问题总是与物理学所称的“无量纲数”联系在一起。我们在第六章中曾简单提到，最好是用比值的形式来表达数量。 113 某位候选人以 9870325 张选票赢得总统大选的说法几乎毫无意义。只有当你发现 9870325 票占全部选票的 87% 时，你才可能会质疑选举结果。无量纲数是一个消除了单位——物理学量纲——的比值，只剩下“纯”数。例如铅的密度是 11 克每立方厘米，或者 0.4 磅每立方英寸。这些数字看起来彼此完全不同，对我们来说也毫无意义。换个说法，铅的密度——无论是英制系统、米制系统，还是 Potrzebie 系统[①]——大约是水的密度的 11 倍。这就是无量纲数。现在我们就可以苹果比苹果，煎饼比煎饼了。

重子－光子比是 10 亿比 1，宇宙学常数要比宇宙暗能量大 120 个数量级，都是无量纲数。说两个质子间的静电力要比它们之间的引力大 10^{36} 倍，也是用无量纲数来描述。

114

当我们提问“为什么”这些数值这么大的同时，我们自然引入了“因为事情就是这样”的回答。我们不应该对这种回应置之不理。另一方面，物理学家认为所有无量纲数的值都应该“自然”地大致相等，最好接近数字 1。如果哪个特别的数值远大于或者远小于其他值，那么就会遇到我们观测到的是精心调制的宇宙那样的问题。最好的办法是找到合理的理由，说明无量纲数为什么会取这样或者那样的值。

在物理学的历史上，“为什么”通常会变为“怎么样”。那么多宇宙学家把宇宙学常数难题视为“宇宙学最重要的问题”，说明他们觉得这事真的很严重。

115

精心调制问题是真的吗？还是只是个哲学问题？

① Potrzebie 系统是漫画家艾尔·卡普虚构的计量系统。——译者注

第九章

星系与我们同在

还有其他问题需要立即关注。第五章中描述的宇宙学原理要求宇宙在足够大的尺度上是均匀同性的。限定语“足够大”显然是故意且便利的模糊表述，但 20 世纪绝大多数的宇宙学计算以简单而非哲学之名，基于宇宙绝对均匀的假设。原初核合成的计算就是经典案例。然而，宇宙并不是均匀的。任何尺度上都不均匀。你可能见到过计算机给出的宇宙大尺度数值模拟结果，如图 11 所示，有着好像肺内结构一样的长纤维，或者是 116

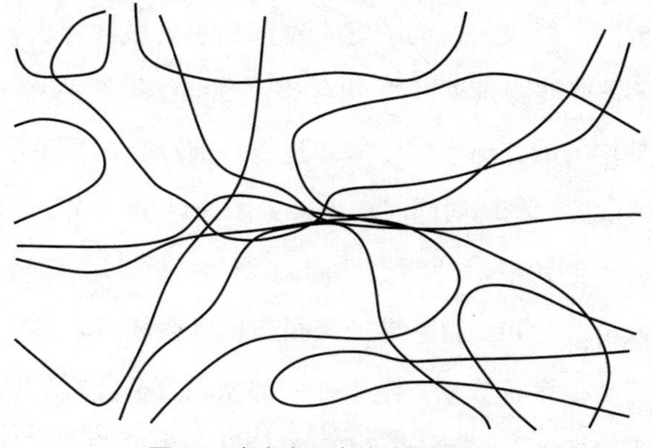

图 11　宇宙大尺度数值模拟结果

杰克逊·波拉克的绘画。

图中的纤维就是超星系团，是可观测宇宙中最大的结构。超星系团能拥有几十万个成员星系，直径可达几亿光年。银河系太小了，在这张素描图中根本不可见。

由于在任何严格的数学框架中，都不能认为超星系团是随机分布的，所以我们就必须面对这个问题：宇宙大尺度结117 构是如何形成的？如果宇宙学原理严格成立，那么根本就不会有这样的网状结构，更不用说我们。宇宙中存在不规则结构的事实要求，均匀的大爆炸模型必须被修改，无论开始时宇宙是多么均匀，它必须迅速演化成非均匀的。而且，现代标准宇宙模型也必须修改，普通物质和辐射要让位给暗物质和暗能量。

✷

在过去的 40 年里，宇宙学研究的主角就是理解宇宙大尺度结构的成因。解决问题的关键一直都是宇宙微波背景辐射。虽然发现宇宙微波背景辐射的 30 年来，观测结果似乎一直显示辐射背景是均匀的，但宇宙学家知道，目前存在的星系必定是与观测到的微波背景辐射同时产生于大爆炸后 38 万年，而星系的诞生一定会在宇宙微波背景上留下微弱的痕迹。

1992 年，COBE 卫星终于探测到这些微弱的痕迹，主流媒体——甚至很多权威宇宙学家——都激动万分，宣称这项发现118 是“上帝的指纹”。老实讲，COBE 团队的的确确交了好运，取得的成果值得开香槟庆祝，但宇宙学家知道如果观测结果是什么都没有，那才更有趣。物理学就在理论和观测的碰撞夹缝中得以存在——一定是有什么或者是什么地方出错了。而COBE的结果只是简单肯定了理论预测。

方便起见，我将把“大尺度结构形成”简称为星系形成，它是宇宙学大统一的最好例证，显示了精确观测、粒子物理和数学是如何共同构筑令人信服的宇宙图景的。

✳

从最简单的层面看，星系形成就是引力与宇宙膨胀对抗的过程。引力试图让物质聚集形成结构；而宇宙膨胀试图阻止结构的形成。谁，或者说是什么，取得胜利了呢？

为了更令人信服地回答这个问题，让我们先聊聊声音。要谈论声音，我们得先说说高卢人。与所有高卢人[①]一样，物理学被分为三个部分：粒子、弦和波。对一位物理学家来说，不是粒子的东西就是弦，如果二者都不是，那就是波。牛顿物理学是有关粒子的；现代场论是有关弦和波的物理学（第八章讲到的真空能就是明证）。一位真正的物理学家立刻就会把任何问题都归结为有关弦的问题，如果需要，再加上波，或者如果讨论的是星系形成，那就是声波和光弦。 119

声波，与除光之外的任何波一样，都是穿过某种介质的扰

① 历史上，高卢人通常被分为三个主要的社会群体：贵族、平民和奴隶。——译者注

动，例如空气。立体声扬声器会振动，扬声器的振动交替压缩前方的空气，使得它膨胀——或者用物理学家的词汇说，变得稀薄。实际上，一小包空气会被压缩，直到气包内的气压大到不能再进一步压缩为止，然后气压使得空气包重新膨胀。当空气包的气压降至低于周围空气气压时，周围空气会再次压缩空气包。空气是一种弦。

扬声器就构建了一系列能够在房间里传播的振动。正是这些振动形成了声波，如图 12 所示，其传播的速度由空气的密度和压力决定。在一个普通房间内，声波传播的速度大约是每秒 340 米。物质密度越大，声波在其中传播的速度越快。例如声音在钢铁中传播的速度可以达到约每秒 6000 米，是在空气中速度的 17 倍。

120

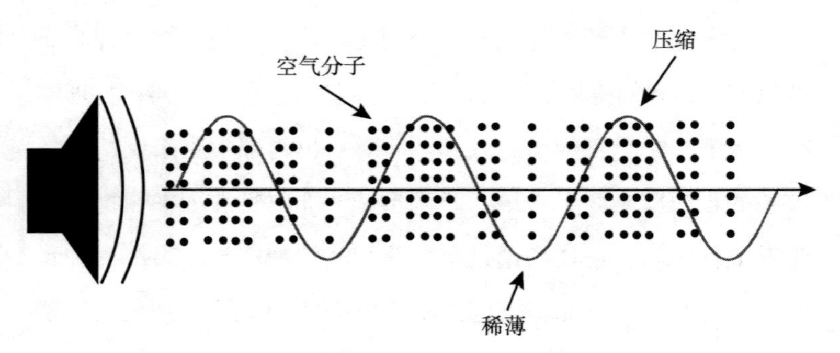

图 12　声波传播过程中空气的变化

在简单的声波中，空气的压力或者是密度会按照正弦曲线从高到低振动，如图 12 所示。任何两个连续的压强峰值或谷值之间的距离就是振动的波长，可被听见的声波波长范围是米级。[①]

现在让我们走到户外。地球大气可以被视为一间大房间，121 如果不是有大气压来抵消地球引力，这个大房间就会因自身的重力而坍塌。而事实上，大气压足以阻止这种灾难的发生。与室内的情形一样，如果大气中一个气柱被压缩一点，那么气压就会增大，使得气柱再次膨胀。膨胀会一直持续到气柱内的压强比周围空气压强低，这时气柱就会再次被压缩。物理学家认为大气处于稳定状态，能够抗衡引力塌缩，只是经历着“声学振荡”——一个复杂的声波专用名词。

但是假如大气层的厚度达到地球直径的1000倍，大气自身的重力就要超过气压所能抗衡的极限，那么大气在引力的作用下就会塌缩，而不再振动。

✷

早期宇宙的情况与地球大气很相似。如果在大爆炸后不久，原初粒子汤能够在整个宇宙中均匀分布，那么物质间的引力作用就可以使得物质开始聚集。早期宇宙中不存在气压，有的是 122 光压。在第五章中我们看到，再复合之前光子在与电子发生碰撞前，不会传播太远。光子撞击物质时就会对其施加压力，这与在太阳系中航行的帆板式飞船的动力是来自太阳的光压是同样的。这种光压抗衡着导致物质的塌缩的引力，继而形成声学

① 见第49页脚注。

振荡，正如空气中的声波那样。

房间里的空气与早期宇宙中的光之间第一个主要区别，是宇宙粒子汤的密度要比空气大得多。钢铁的密度要比空气大，其中的声速是空气中的17倍，但早期宇宙中的声速可以达到光速的60%(精确的数值是$c/\sqrt{3}$)。因此，原初宇宙的“建筑材料”太坚硬了，最小的塌缩结构的质量都要比一个超星系团还要大（一个典型超星系团的发光物质质量大约为10^{16}个太阳质量）。换句话说，在极早期宇宙，不可能形成任何结构。

123　我们还记得在再复合时期，宇宙微波背景辐射出现，那时能够形成中性原子，光子不再与物质粒子发生碰撞。这等于是说作用在物质上的光压降到几乎为0，与此同时，原初粒子汤的密度大大降低，不再那么坚硬了。结果就是允许质量更小的结构塌缩——实际上，质量大约为10^5个太阳质量的结构就可以开始形成，这个数值比银河系质量的百万分之一还小，大约是一个球状星团的质量。

再复合时期光子和物质分道扬镳，在此之前，它们本质上是混成一锅汤的，所以当物质开始聚集时，光子也随着一道聚集。这些细微的光子密度变化，在宇宙微波背景辐射上以微弱的温度变化形式得到复刻。COBE卫星发现的上帝的指纹，其实就是这些细微的温度起伏，COBE的继任者WMAP（威尔金森各向异性微波探测器）以更高的精度证实了这个发现，而更高端的Plank（普朗克）探测器则以惊人的精确性再次肯定了测量结果。虽然微波背景的温度起伏只有十万分之一开，但它们足以产生由于引力塌缩而形成的结构，正像我们今天观测到的那样。如今，“自下而上”的塌缩图景成为广为接受的星系形成理念：最先形成的是最小的结构，然后它们逐渐并合形成更大的结构。甚至在你读到本句话时，超星系团仍在形成。

124

在这张图中是不是遗漏了什么？

125

第十章

宇宙管风琴

前文将房间作为宇宙的类比，忽略了二者之间最本质的差别：宇宙在膨胀。由于膨胀会让结构彼此远离，因此妨碍了引力塌缩。膨胀与塌缩之争的结果，取决于具体的膨胀速度，而膨胀速度又依赖于宇宙中有什么成分以及有多少。

光子的行为与物质不同，暗能量则与二者都不同，因此一点也不必惊讶，宇宙的膨胀速度不仅与其成分的密度有关，而且与成分的本身属性有关。一个由可见或者不可见物质构成的宇宙（物质主导的宇宙，第五章），其膨胀速度是持续变慢的。而一个由辐射主导的宇宙，其中光子或者中微子占多数，膨胀的速度也会变慢，但变慢的速度不同。充满暗能量的宇宙——由宇宙学常数主导——膨胀的速度则维持不变。高度弯曲的宇宙则拥有完全不同的演化方式。

因为宇宙膨胀速度完全取决于宇宙的成分，很自然的，你会认为改变宇宙成分就会改变星系形成的图景和结果。事实确实如此。而且我们还很幸运，因为宇宙学家能够排除大多数可能。接下来的问题是：要允许星系出现在目前的宇宙纪年中，宇宙物质的精确配方到底是什么呢？

✷

为了回答这个问题，让我们再回到关于声音的类比，这次是管风琴。教堂管风琴最显著的特征就是排布的数百根不同长度的管子。管子的长度决定了音调的高低。具体来说，管子的 127 长度精确地确定了管内发生共振的波长或者是频率。管风琴的管子各式各样，但最基本的样子就是顶端和底端都敞口。这样当声波穿过管子压缩空气时，使其变得稀薄，而两端的压力则保持与室内空气压力相同。这就创造了空气能够在管内共振的条件。正如图 13、图 14 所示，在这样的空腔内能够满足条件的最长波长是管子长度的 2 倍。这就是我们听到的第一和声或称基础和声。

波长恰好等于管风琴管子长度的声波也满足共振条件。因为其波长是基波的一半，因此它的频率就相应的是基波的 2 倍。这是我们听到的第一个泛音，或者称为第二和声。第三个和声的振动频率是基波的 3 倍，也符合共振条件，其他更高频率的声波以此类推。在所有情形中，从压强最大或最小值处到最近的室内压强处的距离，都是四分之一波长，或者称为四分之一振动。

从根本上说，宇宙就是一架管风琴。 128

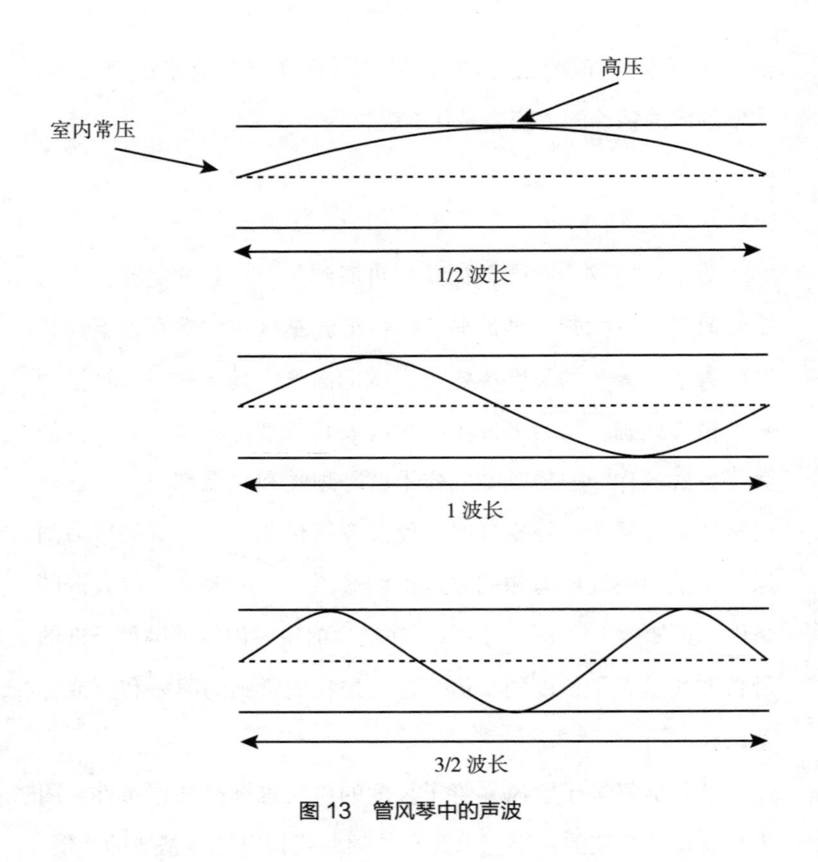

图 13　管风琴中的声波

✷

如果我们把一架管风琴产生的声波描绘出来，其样子一定比简单的正弦曲线要复杂得多，但是正像图 14 左图那样，是对其波形理想样式的描绘。

现在你已经知道，乐器演奏的音调是由基础和声加上所有更高频率的和声构成的。因此我们就可以认为无论什么音调，

129

都是以基础和声（基音）为基础，加上泛音组成，如图 14 右图所示。每个频率处声音的强度决定了原始音调的形状。用数学语言来说，将音调分解为和声或者说泛音（谐波）的技术称为频谱分析。将已经分解为不同泛音的波描绘出来，就是图 15 所示的那样，标明了各频率处声音的能量大小。这就是声谱——与光谱和热谱一样。这些图给出的是简单的示例，只包含了三个谐波。

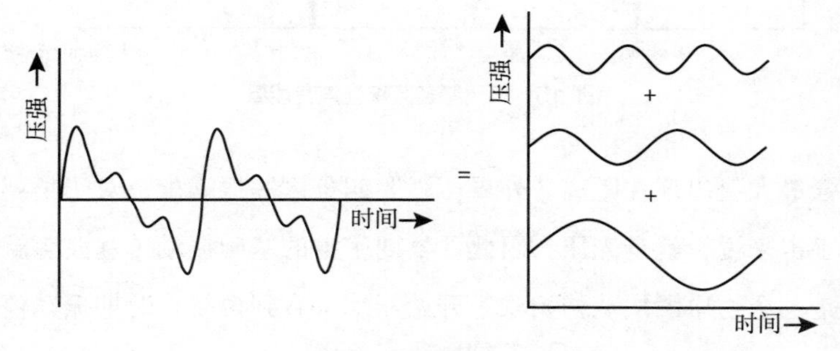

图 14 基音与泛音

早期宇宙就是一架能想象得到的最宏大的管风琴。要记得在宇宙微波背景上探测到的温度起伏，实际上是宇宙物质密度涨落的表达。这些涨落幅度并不一致。普朗克空间望远镜绘制的精细图像显示，某些区域有更高的密度，密度涨落的谱形与一架管风琴的声波谱形完全类似。 130

实际上，密度团块的物理尺度完全由早期宇宙的共振频率决定。想象中大爆炸之后不久，所有的物质都均匀分布。然后物质开始聚集，但光压会让团块发生振动。直至再复合时期，光子与物质脱耦，这种振动才会停下来。对于管风琴来说，压 131

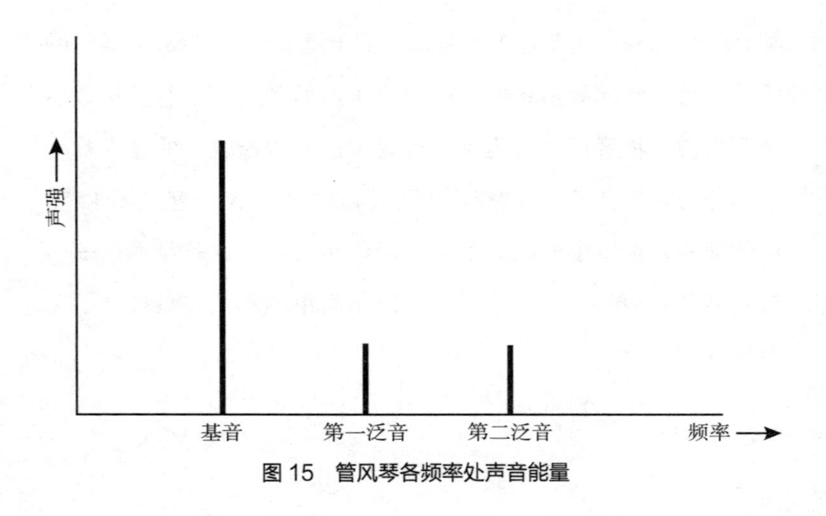

图 15　管风琴各频率处声音能量

强最大值出现在距离“外界压强”四分之一振动处，对于早期宇宙来说，就是光压。因此，早期宇宙的基础振动（基波）就是一个物质团块从初始状态开始压缩，直到再复合时期振动停止时，刚好完成一次振动。第一个泛音则完成一次压缩一次膨胀。第二个泛音对应着物质团块完成一次压缩，一次膨胀，再一次压缩。

你可能会提出反对意见，因为管风琴的管子是有物理长度的，而这里我谈论的是时间——大爆炸和再复合时期之间的那段时间。但每一段时间都对应着一个长度。这里，长度就等于从大爆炸到再复合这段时间内声波传播的距离。因为此时的声速大约为 0.6 倍的光速，因此对应的距离大约为几十万光年。与管风琴的管长类似，涨落的基波波长是这个距离的 4 倍。对应的一系列泛音（谐波）的波长则按比例减小。

自这些振动将自己的印记刻在宇宙微波背景上以来，宇宙大约膨胀了 1000 倍。由于波会随着宇宙一起膨胀，因此所有的谐波波长都会被等量拉伸，但它们可以转为今天我们看到的深空中的角距离。其中基波对应的角距离应该是大约 1 度——相当于满月视直径的 2 倍。其他谐波对应的角距离均相应减小。 132

最令人惊奇的是，长达几十年的地基和卫星观测结果显示，预言的所有谐波均被探测到。例如，普朗克空间望远镜的观测结果显示，原初密度涨落可以被分解为声谱。这种声谱被称为重子声波振荡，对大多数人来说就是一种声波，对于“铁粉”来说就是上帝的指纹，而这张声波图形则出现在所有宇宙学研讨会上。图 16 第一个峰值就代表宇宙管风琴的基调，其他的峰值则对应其他的泛音。

因为物质团块的聚集依赖于宇宙膨胀速度，而后者又取决

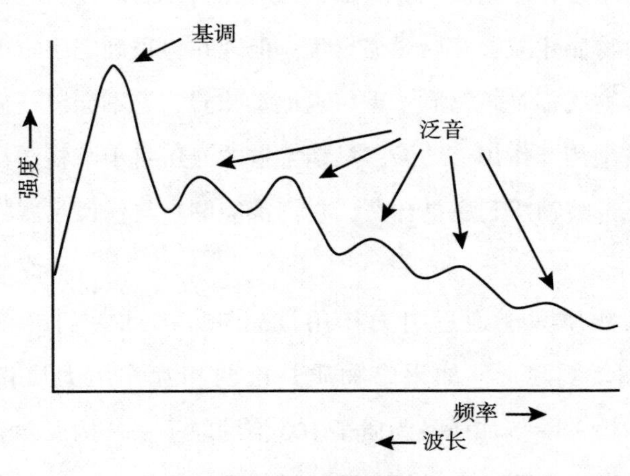

图 16　宇宙管风琴

于宇宙的成分，所以这张草图应该反映这个事实。实际上，宇宙微波背景辐射涨落的波谱已经成为对宇宙学模型最灵敏的检 133 测。在一个闭合宇宙中——像球面一样弯曲的宇宙——遥远天体看起来要比在平直宇宙中更大。这会造成波谱上的峰值都朝更大的角尺度方向移动，对应的是上图的左侧。由于模型预言的峰值位置与观测到的恰好吻合，因此宇宙必定是平直的，任何人都看得出来。这就是我在第三章中声明宇宙几乎是欧几里得式的（平直的）根本原因。

如果宇宙是平直的，那么按照定义，宇宙中所有成分—— 134 普通物质、辐射、暗物质和暗能量——的密度总和，一定等于第四章中讨论过的临界密度。事实正是如此，宇宙戏法就是平衡宇宙中所有的成分，给出与观测结果最符合的配方。

让我们看看物质的情况。如果普通重子物质（中子和质子）是宇宙中仅存的物质成分，那么只有在再复合时期，光压完全消失，物质才能够开始塌缩聚集。但现在你很确定宇宙中的绝大部分物质都是暗物质，更精确地意味着它们不会以任何方式与光发生相互作用。结果，早期宇宙的光压对于暗物质没有影响，无论暗物质到底是什么，它们都能够参与任何尺度的声学振荡。

暗物质只能通过引力作用被感知，因此它们自然会聚集塌缩。实际上，如果暗物质是由弱相互作用大质量粒子（WIMPs）——例如质量为质子100倍的粒子——构成，那么几乎在大爆炸刚刚结束时就会开始聚集塌缩。因为在宇宙变成物质主导时，正如第五章中描述的那样，早于再复合时期，暗物质的存在变得相当可观，它们就提供了推动重子物质聚集的引力核心。更多物质的聚集转变成原初声谱中更高的峰值。 135

另一种情况，假如暗物质是由中微子构成的。因为暗物质就是暗物质，这种情况下，中微子和弱相互作用大质量粒子没有区别，除了我们确知中微子是真实存在的。所以，中微子同样可以提供让重子物质开始聚集的引力核心。问题是，与弱相互作用大质量粒子相比，中微子太轻了，其在早期宇宙中自由运动的速度接近光速。这样的速度太快，中微子无法在自身引力作用下聚集，除非它们的质量能达到差不多超星系团的质量——在那种情况下，引力核心的尺度接近整个宇宙的尺度，根本不会形成球状星团那样的小尺度结构。

弱相互作用大质量粒子质量更大、速度更慢，这样的粒子被称为冷暗物质，相对的，速度更快的粒子被称为热暗物质。总体而言，声谱中频率更高的泛音，代表着更小尺度的物质聚集，它们在热暗物质宇宙模型中都被抹除了。由于宇宙中确实存在更高阶的谐波（泛音），因此宇宙学家认为宇宙中的暗物质是冷的。 136

宇宙学常数是决定今天宇宙膨胀速度的主要因素，但对宇宙微波背景辐射谱的影响并不大。虽然在今天的宇宙能量密度中，宇宙学常数的权重“远超过”物质（可见物质和暗物质），但在早期宇宙，它与物质的密度是相同的——毕竟，宇宙学常数是个常数。物质和辐射的密度在过去迅速增大，仅在几十亿年前，就超过宇宙学常数的能量密度。所以宇宙学常数对于宇宙微波背景辐射的形成几乎没什么贡献，因为那个时期远比几十亿年要早得多。然而，由于宇宙在加速膨胀，以及至今我还没有提及的其他原因，宇宙学家相信宇宙学常数确实存在。

原因之一就是宇宙微波背景引力透镜。正如第七章中我们讨论过的，晕族大质量致密天体（MACHOs）能够扭曲位于其后方光源的像，普朗克空间望远镜给出的宇宙微波背景辐射图像也会被任何位于我们和可观测宇宙边缘（宇宙微波背景辐射诞生之地，距我们接近 140 亿光年）之间的物质——例如超星系团——扭曲。就像放大镜所产生的像与其在眼睛和成像物体之间的位置有关一样，CMBR 图像的扭曲取决于作为透镜的物质所处的位置。在膨胀宇宙中，这取决于上述所有因素，包括宇宙学常数。要获得与观测到的宇宙微波背景辐射谱最吻合的结果，宇宙成分中就必须有暗能量。

137

这样，我们终于得到了今天标准宇宙模型的配方，它通常被缩写为 ΛCDM，即兰姆达冷暗物质。对观测结果最好的拟合曲线要求宇宙中 68.5% 的成分为暗能量，26.7% 的成分是暗物质，4.8% 是普通物质——但请不要以此为正式结论引用。

✷

虽然 ΛCDM 模型非常成功，但它的确留下很多开放性的问题。首先，一旦掌握了所有的宇宙成分，计算出今天的哈勃常数值就应该易如反掌了。不幸的是，通过重子声学振荡和引力透镜计算出的哈勃常数值是 67.4[1]，这里使用的是天文学家采用的标准单位，而利用超新星测量结果给出的数值则是 73.9，二者的差距达到 10%。天文学家对于哈勃常数的执着堪比十字军，所以显然他们不会轻易放手，直到问题得到彻底解决。 138

10% 的差别很重要吗？观测到的对哈勃常数小小的偏离确实导致了宇宙加速膨胀的发现。然而目前的情况，更可能是计算中哪里搞错了。很快，测量结果就会达到某个点——例如，假设是 1% 的差别——那时更精准的对于哈勃常数值的测量将不再能引导我们发现新物理，所以在那个时刻到来之前，思索我们为什么要精确测量哈勃常数才是明智的。

更重要的是，我还没有真正讨论结构形成，而只是讨论了结构形成的开端。随着宇宙演化，星系和恒星开始形成，涉及的物理变得越来越复杂，因为引力之外的其他力开始起作用了。准确来说，在宇宙微波背景辐射产生后大约几亿年时间里，宇宙进入“黑暗时代”。“黑暗时代”结束时，最早的星系开始形成。之后的几亿年时间，星系开始聚集为星系团，而今天的宇宙中超星系团仍在形成。 139

在宇宙年龄之内能够出现所有这些结构，说明在宇宙微波背景形成时，上帝指纹的尺度必须与观测到的一致：十万分之一的密度涨落。

更有意思的是，上帝的指纹谱还有个有趣的性质，宇宙学

① 天文学家通常写为 67.4km/s/Mpc，千米每秒每兆秒差距。

家称之为尺度无关。简单来说，尺度无关的意思是所有事物在任何尺度上看起来都一样。用放大镜观察一片蕨类植物的叶子时，你能看到小尺度与大尺度下看到的结构是相似的。“蓝多湖”黄油的包装盒图案就是一位美国女性原住民拿着一盒“蓝多湖”黄油，这个包装盒上又是一位美国女性原住民拿着“蓝多湖”黄油……如果一架管风琴波谱中每个八度音阶的声强都是相同的，我们就可以说它的波谱是尺度无关的。如果你愿意，140 可以称其为“蓝多湖波谱”。①

早期宇宙中，相对于物质聚集的空间而言，物质聚集的强度为常数。很难理解为什么重子声学振荡产生的谱是尺度无关谱，但它就是这样的。

141 是什么决定了上帝指纹的大小和谱形？

① 更精确的定义是每立方波长每个八度的声强都是常数。对于宇宙微波背景辐射来说，“强度”指密度扰动幅度的平方值。

第十一章

第一次闪烁：宇宙暴胀

至此，宇宙的故事讲述到大爆炸之后 0.0001 秒，原初核合成马上就要开始。很自然的，我们会想知道更早时期的宇宙发生了什么样的事件，但事情开始变成更多猜测的成分。让时间回溯到大爆炸之后大约 0.000001 秒（1 微秒），我们推测此时中子和质子因极端高温而被分解为夸克（夸克是中子和质子的组成成分），这一点最近已经被地面的粒子对撞机实验所证实，但在宇宙更早时期，是否会出现过多的新粒子仍不可知。大爆炸后十亿分之一秒，应该能看到希格斯玻色子的身影。希格斯玻色子就是传说中能够帮助其他粒子获得质量的粒子，此处我仅是顺便提到它，因为在整个宇宙图景中，希格斯玻色子并非主角。显然，有关奇点的可怕想法已经跃跃欲试了，在时间 t=0 时，一切都将灰飞烟灭，但眼下让我们继续回避与奇点的直接遭遇，而是像宇宙学家那样，只从大爆炸之后的瞬间开始思考，尽管有着那么多的不确定性。

142

1980 年刚过，一个有关大爆炸之后 10^{-32} 秒的新理论引起了宇宙学界的注意——很快，也激发了公众的想象。新理论被命名为“暴胀”，是该理论的倡导者阿兰·古斯创造的，其意义显而易见。尽管古斯一直在各种学术研讨会上宣讲暴胀理论，但类似的想法已经由美国的狄摩西尼·卡扎纳斯和苏联的阿列克谢·斯塔罗宾斯基发表过。

由于很多原因，尤其是理论的名字，暴胀理论一飞冲天。它几乎立刻就被纳入标准宇宙学模型，教科书把它当作板上钉钉的理论介绍，40年来，暴胀理论一直充当着宇宙学研究的基石。你应该知道，暴胀理论不是像量子力学那样的标准理论，后者已被无数实验和观测证实。实际上，到目前为止，暴胀理论给出的是上百个模型，其本意都是解释大爆炸理论的“缺陷”，正如我之前讲到的。这些“缺陷”并非观测结果的异常，而是理论或者说是哲学上的难题，是标准大爆炸理论无法解释的。它们更像是第六章提到的光子-重子比的困惑，或者是第八章所说的宇宙学常数问题，而不是像水星近日点进动这样的麻烦。暴胀理论是否真能解决这些问题已经越来越成为热议的焦点，它究竟是胜利者还是会被扫进历史的垃圾堆，那就是未来宇宙学家要做出的决定了。

143

*

罗伯特·迪克一直强调，暴胀理论要解决两个问题，首先是平直问题。从始至终，本书谈及的真实宇宙都几乎是平直的，这一点已经被观测所证明。那么为什么宇宙是平直的？

144

“为什么不是呢？”你可能会这样回应，但事情不会这么简单就被解决。如果现在宇宙是近乎平直的，那么宇宙密度就一定接近临界密度，正如第四章所述，临界密度是区分“闭合”宇宙和“开放”宇宙的临界值。这种情况的可能性有多大？为了说明，假设今天宇宙的密度值为临界密度的99.5%。那么很容易计算出在大爆炸之后1秒，也就是元素开始形成时，宇宙密度的起伏必须在临界值的10^{17}分之一范围内，而在大爆炸后10^{-36}秒时（这个时间不是我随便选定的），宇宙密度的起伏则不能超过临界值的10^{52}分之一。换句话说，宇宙必须被以不可想象的精度所调和，才能保证今天的平直性。

即使是那些愿意相信偶然巧合的人，也会认为大爆炸理论是不可能产生如此平直的宇宙的。至于光子－重子比和宇宙学常数难题，则更像“为什么”类型的问题。与之前一样，宇宙学家发现把“为什么”类型的问题转化为“怎么样”类型的问题更合适——他们更倾向于避免任何精细调制宇宙，而是要找出某种机制能够让宇宙变得平直，不管它到底是如何开始的。 145

但是当我们只有一个单一宇宙时，“可能”和“不可能”又到底意味着什么呢？此处我们迎头撞上了单一宇宙的难题，让我们在下一章再来对付它。

✷

迪克宣称的暴胀理论要解决的第二个难题就是视界问题。观测到的宇宙微波背景辐射温度在所有方向都异乎寻常的均匀。

甚至前面章节所称的“上帝的指纹”对于均匀性的改变，也只相当于对世界第一高建筑物迪拜哈利法塔，改变一个玻璃球直径的高度。这种异乎寻常的均匀性从何而来？是否是另一个巧合？

为了让情况更生动，想象一下，假设可观测宇宙中有 10^{87} 个光子，这是个相当大的数。由于这些光子都处于可观测宇宙内，也就是说它们都在自大爆炸以后光能够传播的距离范围内——第四章讨论的宇宙学视界。因为任何信号传播的速度都不会超过光速，因此宇宙学视界就给定了信息交换的终值边界：

146 如果两个物体的位置超出彼此的视界，那么它们就不会对彼此造成任何影响。如图 17 所示，A 点的视界是自大爆炸起光所传播的距离，（光速）×（宇宙年龄）=ct。A 和 B 都位于这个距离之内，彼此可以有影响。A 和 C 则彼此无法联系，直到 A 的视界增长到等于 A 和 C 之间的距离为止。A 和 B 被称为有因果联系，

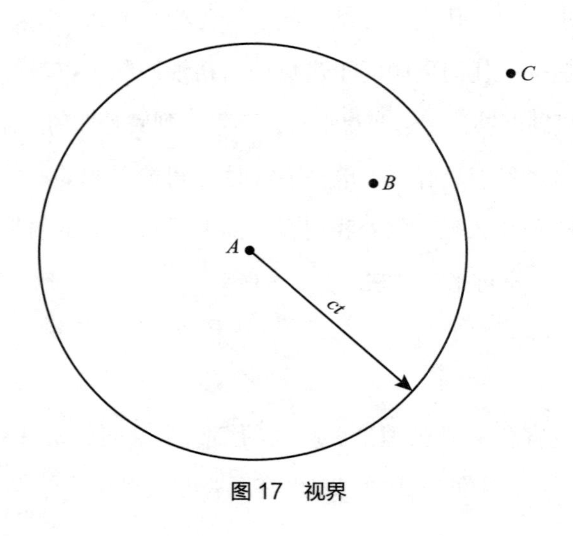

图 17　视界

而A和C则没有。

按照定义，任何处于今天可观测宇宙之内的事物，都处于宇宙学视界内。同样按照定义，视界以光速增大，因此朝大爆炸回溯时，视界也以光速收缩。另一方面，宇宙的膨胀速度——星系彼此远离的速度——比光速小。因此，回溯过去，宇宙收缩的速度要比视界收缩的速度小。结果当我们接近大爆炸时，视界内的宇宙只占据比成为今天可观测宇宙更小的部分。在宇宙微波背景辐射产生之时，位于视界内的宇宙只相当于今天宇宙的十万分之一——也就是说只有 10^{82} 个光子。

147

这意味着宇宙微波背景辐射中位于两个相距遥远区域的光子，在宇宙微波背景产生时彼此无法联系。就像图中的A和C，它们还没有因果联系。那么它们又是如何变得具有一模一样温度的呢？这就是大爆炸的视界问题。

✷

暴胀理论宣称能解决的第三个问题是磁单极问题。根据大统一理论（GUT），强核力、弱核力以及电磁力在大爆炸后 10^{-37} 秒、宇宙温度为 10^{29}K时，可以被统一为一个“场”。随着宇宙膨胀，这个统一场分化为不同的场，就产生了所谓的磁单极。磁单极是指完全孤立的磁北极或者磁南极，就好像电场的正电荷和负电荷一样。但是尽管自然界里到处都是孤立的正电荷和负电荷，例如质子和电子，但从没有人发现过孤立的磁北极或者磁南极。所有的磁铁都同时拥有北极和南极，如果把一块磁

148

铁从中间分成两半，也只是得到两块更小的磁铁，每一块都有自己的磁北极和磁南极。

然而，有些大统一理论预言早期宇宙中应该产生过大量的磁单极，而且它们的质量非常大（比质子质量大 10^{16} 倍），应该完全主导那时的宇宙密度。这就是磁单极问题。

✷

暴胀理论对以上三个难题的解答十分优雅简洁，普通物理 149 学家都能理解。暴胀理论假设在大统一时代结束时（大爆炸后 10^{-36} 秒至 10^{-32} 秒之间），宇宙经历了一次巨大的指数膨胀，在短得不可想象的瞬间，尺寸增大了 10^{27} 或者 10^{28} 倍。这相当于一粒爆米花核瞬间膨胀至整个可观测宇宙的大小。

如果你是一只在爆米花核表面爬行的小蚂蚁，那么当它的大小瞬间增大 10^{27} 倍时，其表面就会看起来异常平直。这就是暴胀理论对于平直难题的解答。

磁单极问题也与平直问题一并解决。宇宙中数量众多的磁单极只是简单地被巨大的膨胀稀释了，其密度降至大约每个可观测宇宙中只有 1 个磁单极，而我们还没有找到它。

视界问题则复杂一些。问题关心的是天空中两个相距遥远的部分是如何彼此联系、抹平差异从而产生异常均匀的微波背景辐射的。由于在标准宇宙学模型里，视界朝向过去收缩的速度要比宇宙收缩的速度快，因此在大爆炸后 10^{-36} 秒时，视界的 150 大小要比当时的宇宙小 10^{-27} 倍。所以，事实上宇宙中任何粒子之间都没有联系。但另一方面，按照定义，位于视界之内的粒子是可以相互联系的。如果那部分区域因暴胀而膨胀 10^{27} 倍，那么正好就是今天可观测宇宙的大小。

这就是暴胀理论宣称的结果：假设今天的宇宙是从一个爆米花核大小的区域产生的，那其中的光子已经彼此产生联系并平滑掉了任何涨落；这一小块区域的暴胀，就产生了均匀的宇宙微波背景辐射。然而必须指出，暴胀本身并没有解释平滑是如何发生的；它只提供了平滑产生的必要条件。

*

暴胀理论变得如此流行的根本原因是上帝的指纹，而与上述三个难题无关。宇宙微波背景上的涨落，意味着相对于 2.7 开的背景来说，温度的变化只有 1/10 万开。而且微波背景还是尺度无关谱。这两个特征都是观测结果。它们是如何产生的？ 151

早期暴胀模型能够给出解答。第八章中我们说过，物理学家认为真空里充满了很小的能量涨落，就是所谓的真空能量涨落。暴胀理论假设这些量子涨落在大爆炸后立即出现，产生于所谓的量子引力时期，我们将在第十四章中讨论。暴胀带着涨落一起，直到它们成为宇宙微波背景辐射中看到的涨落。而且，这样的暴胀使得振动声谱必然是尺度无关的。

*

所以如果暴胀确有其事，那么看起来它能够对我们宇宙某些令人困惑的特征做出解释。但是，暴胀本身又是如何产生的呢？几百种不同的暴胀理论都给出不同的答案。大多数理论假定存在一个新的场，有点像暗能量。要记得宇宙膨胀速度依赖于其物质能量成分。如果宇宙由暗能量——宇宙学常数——主导，那么弗里德曼的方程告诉我们，宇宙会随时间以指数方式膨胀。实际上，因为今天的宇宙由宇宙学常数主导，所以它正
152 在以近乎指数方式膨胀。

在暴胀图景中，大爆炸后 10^{-36} 秒至 10^{-32} 秒之间发生的事情大同小异。此时宇宙由一种新的能量主导，这种能量不必然是今天的暗能量，但在一段时间内会表现得与宇宙学常数相似，如图 18 所示。这种近乎为常数的能量产生了指数形式的膨胀，在暴胀结束阶段，能量衰减直至消失。这张图被称为势能图。如你所知，任何系统，例如位于山顶的皮球，都会趋于寻找最低势能，正因如此，山顶的皮球才会滚落。物理学家经常把宇宙视为位于能量曲线顶峰的皮球，而能量则由暴胀场提供。随着皮球沿着几乎平直的山坡下滑，暴胀开始了。在暴胀终期，皮球迅速掉入井里，损失了所有能量。

然而，物理学家坚持著名的能量守恒定律，不愿意相信作为宇宙主导的能量会不留痕迹地消失。于是，基本的图景就是
153 这样：在暴胀时期宇宙膨胀得足以解决那些宇宙学难题。巨大的膨胀同时也彻底清空了宇宙的物质成分——磁单极、光子、中微子和任何其他物质。当暴胀结束时，驱使暴胀产生的场衰减消失，将其能量转化到组成今天宇宙的粒子中。暴胀，加上

随后的所谓“再加热”时期，全部都发生在远远短于一次眨眼的时间内。

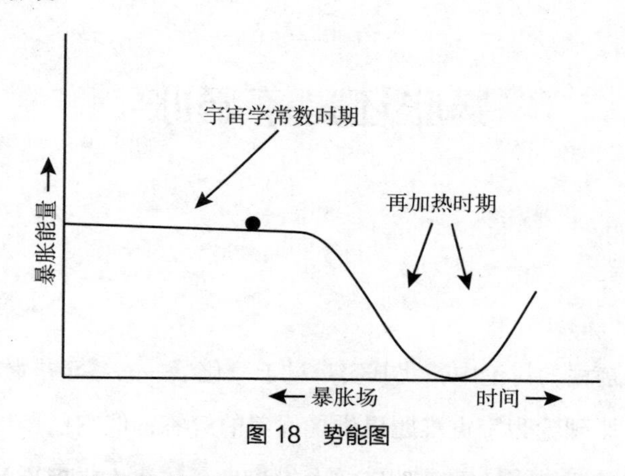

图 18　势能图

为什么暴胀的能量会消失呢？最初的理论是基于著名的相变现象。如果缓慢而且小心地冷却水，那么水的温度可以降至远低于冰点，但如果有一粒尘埃粒子混入水中，它就会成为“成核中心”，四面八方的水都会迅速结冰。在大统一理论中，当统一的力分裂成不同部分时，早期宇宙空间的真空能量也会发生类似的相变。在暴胀期间，很大区域的真空能开始变得“过冷”，最终经历相变成为今天的宇宙。后期的暴胀理论版本只是假设一种新的场，其势能图与图 18 非常相似。　154

概括而言，本章描述的图景阐明了暴胀理论是如何被假定可以治愈宇宙头疼的。

是不是遗漏了什么？　155

第十二章

暴胀还是不暴胀

在前面的讨论中，我其实是过于简化了——甚至是撒谎了。虽然暴胀理论为宇宙难题提供了优美的解答，但它已经受到越来越多的质疑，这在科学研究中很常见，如今暴胀理论的前景已经今非昔比，远不如刚刚被提出时那么明朗。

先看看磁单极问题。尽管已努力几十年，但还没有发现任何支持大统一理论的实验证据。也许问题很简单，就是预言大量磁单极存在的理论是错的，这样的话，磁单极问题就不攻自破了。

156

再来看上帝的指纹。大多数网络账号，既包括普通的也包括科技的，都关注宇宙微波背景辐射谱的涨落，以及背景谱是如何与最简单的暴胀预言相吻合的。而且涨落的幅度——1/10 万开——也必须得到合理的解释。很久以来我们就知道，要在最简单的模型中重现这个数字，就必须非常精细地调整第 109 页势能图的形状——如果变化超过 10^{14} 分之一，你可能就得到错误的答案。这是另一个精细调制的例子，迫使我们追问势能是否必须有着如此的形状，这样做，只是用另一个精细调制的问题代替而已。

更进一步，尽管尺度无关谱可能与暴胀理论预言相符，但暴胀并不是唯一能产生尺度无关谱的过程（详见下一章）。如果事实如此，那么又是怎样在众多模型中作选择呢？实际上，暴胀并没有预言一个尺度无关谱，而只是近乎尺度无关。至少已经有一些宇宙学家认为普朗克卫星数据与暴胀理论真正的预言相悖，因此根据观测数据，暴胀理论应该被舍弃，第十三章中将要讨论的模型才是正道。不用说，暴胀理论的支持者们一定是反对的。 157

✷

暴胀图景还向宇宙学家提出了另一些困惑和难题。例如，200 多年前，人们就知道光被窗玻璃反射后会发生偏振。这到底是什么意思呢？光作为一种电磁波，由彼此成直角的振动的电场和磁场组成，这样波才能传播。电场指向的方向被定义为偏振的方向，或者称为偏振的轴。从炽热的灯泡发出的白光是无偏振的，也就是说灯泡发出的光，其电场随机指向所有方向。无偏振光可以被想象为由两束独立的光构成，两束光的电场彼此垂直。当这样的光射到窗户上时，其中一个方向的光会被玻璃优先反射，从而变成偏振光——其电场只在一个方向上振动。

你当然知道这是事实。偏振太阳镜的工作原理就是镜片材料分子按要求排列，只能传播一个方向上的偏振光，这样就把 158

非偏振光的强度降低了一半。因为汽车挡风玻璃反射的光已经是偏振的，所以如果你旋转太阳镜直到它的偏振轴与光的电场方向垂直，你就几乎什么都看不到了。

宇宙背景辐射就是一面硕大的汽车挡风玻璃。在宇宙微波背景辐射产生时，光子撞击电子，使其沿着光的电场方向振动。由于碰撞电子会优先在一个方向再发光，所以光是偏振的。如果原初宇宙粒子汤完全均匀，光子会在各个方向以同样频次撞击电子，因此整体的偏振为 0。但由于精细的上帝指纹的存在，说明宇宙微波背景辐射这面硕大的挡风玻璃并不是绝对均匀的——这就产生了极小的净偏振。

宇宙微波背景的偏振已经被众多极其灵敏的望远镜精确测量过——望远镜实在是太多了，无法一一列举，望远镜的首字母缩写都类似于 DASI（度角尺度干涉仪，位于南极）和 ACT（阿塔卡马宇宙学望远镜，位于智利的阿塔卡马沙漠）——所有观测

159

结果都证实了上述图景。

现在，暴胀理论也预言了原初引力波的存在，那是由极早期宇宙的量子场涨落产生的。回到第三章，我们曾讨论过引力波，它们能穿越时空，像潮汐那样拉伸压缩任何为了测量它们而建立的探测设备。同样，引力波也会作用于宇宙微波背景辐射产生时的原初宇宙粒子汤，造成的不规则起伏也会产生偏振光。然而由引力波造成的对宇宙微波背景的拉伸和压缩，其留下的痕迹与声学涨落所产生的物质聚集（第十章中讨论的聚集）痕迹不同。原则上，灵敏度足够高的望远镜是可以区分这两种

不同的偏振图样的。

原初引力波造成的宇宙微波背景辐射的偏振比声学涨落造成的偏振要小得多，但有些宇宙学家仍然认为如果这样的偏振能够被发现，那么无疑会为暴胀理论提供确凿无疑的证据。尽管 2014 年 BICEP2（宇宙河外偏振背景成像）团队在哈佛大学公开宣布发现了这种偏振，但他们最终又撤回了结果，至今为止还没有发现原初引力波。正如已经提到的，有些宇宙学家认为普朗克卫星的数据已经排除了暴胀理论。 160

✷

然而对暴胀理论的反对之声，主要还是来自其基本假设。虽然我已经多次提及量子涨落，也说明了暴胀理论是如何起作用的，但非常重要的一点是，有关宇宙开端的量子理论尚不存在。暴胀理论并不是真正的宇宙量子理论；而且，暴胀模型只是利用普通的经典物理来“模仿”假设的量子行为。实际上，对暴胀模型的最主要反对之声，就是整个领域完全只是为了产生暴胀这一目的而引入，没有任何观测或者理论证据。

一个与此相关的困难是暴胀模型试图拉伸假想的原初量子涨落振动，直到它们变成在宇宙微波背景辐射中观测到的起伏。但还没有一种机制能够让量子理论转化为经典理论。事实上，如果暴胀持续的时间比解决宇宙学难题所需要的稍微长那么一点点，那么暴胀开始时的振动波长要比 10^{-33} 厘米还短。这是 161 个非常小的数值，物理学家称之为普朗克长度，在这个尺度上，经典物理理论会全部失效；小于这个尺度，应该由量子引力理论接手。因为这样的理论还没有建立，所以我们必须以怀疑的态度来对待任何声称是在量子引力时期可能会发生的事情。

暂且假定暴胀模型合理再现了量子行为。量子场的涨落随机分布在整个宇宙中。小幅度的涨落远远多于大幅度的涨落，尽管如此，偶尔也会发生大幅度的涨落。在暴胀时期，宇宙某处一个大的涨落可能会将第 109 页图 18 中的曲线推高，导致在暴胀结束前，那个区域发生更多的暴胀。随着这个“大泡泡”膨胀，会产生更多的涨落，然后又产生更多暴胀的子泡泡，无穷无尽。暴胀是永恒的，真正意义上的永恒。结果就是个非常不规则的状况，不同子宇宙中有着不同数量的暴胀。在其中一些地方，暴胀可能解决了宇宙学难题，而在另一些地方则没有。这样的多重宇宙似乎是暴胀模型不可避免的结果，我们将在第十五章中更详细地讨论。

162

现在，重要的一点是，虽然多重宇宙在公众中间极为流行，但它提出了观念上的极端难题。让我们来估计一下能够解决宇宙学难题的给定宇宙出现的概率。如果我们面对的是无限多的宇宙，往轻里说，也是很棘手的。当我们随机地把飞镖投向一个 25% 为黄色、75% 为黑色的镖盘时，直觉告诉我们，命中黑色部分的机会是黄色部分的 3 倍。即使面对的是一个无限大的镖盘，我们也会认为打中黑色的次数是黄色的 3 倍，实际上我们可以就这样给出概率的定义。

另一方面，如果镖盘上有无限多种颜色，那么命中任何一

个的概率就是 0。假设有无限多的绿色代表暴胀能够说得通的所 163
有条件，但还有无限多的红色、黄色、黄绿色等。此时命中绿色的概率会比 0 更高吗？与黑黄镖盘的情形一样，我们必须给出类似“在有限大小的镖盘上，命中绿色的概率是命中紫色的 3 倍”这样的限定，然后假定这种描述对于无限大的镖盘仍然成立。

正是暴胀将我们带入这样两难的境地。如果你要问产生能够解决宇宙学难题的宇宙的概率，你必须先确定哪些条件——颜色——要比其他的更有可能，显然还没有达成简单的一致意见。宇宙学家加里·吉本斯和尼尔·图罗克认为，大多数宇宙都没有暴胀到足以解决宇宙学难题的程度。数学家罗杰·彭罗斯则更前进了一步。有关暴胀的方程与牛顿方程非常相似，如果你知道现在的情况，那么就能预言未来或者重构过去。如果你假想一个今天看起来非常不规则而且弯曲的宇宙——远比观测允许的更加不规则、更加弯曲——并且把方程推演至暴胀之前，那么你就得到了一系列约束条件，按照你自己的构想，说 164
明暴胀并不能平滑掉不规则也不能产生平直的宇宙。而且彭罗斯还指出，出乎意料的是，这样不规则的初始条件要比平滑的初始条件更有可能，因此他给出结论——暴胀并不能产生与我们自己的宇宙相似的宇宙。

✷

另一套对于宇宙学难题的解答也经常被提及。有人认为只有接近平直的宇宙才能允许生命演化。如果宇宙过于闭合，它们几乎立刻就会再塌缩而形成大收缩，导致在星系有机会形成之前有一段极为漫长的时间。如果宇宙过于开放，星系也不可能形成。因此，在多重宇宙产生的所有可能性中，我们观测到的宇宙必然是这样，因为毫无疑问，我们就在这里。这是人择原理的另一个例子（我们将在第十五章中进一步阐述）。物理学家对这种言论表示怀疑，因为没有办法能验证它们，但这样的言论会为解决暴胀模型引入的极端难题带来一点曙光，毕竟这样我们只有一个宇宙。

165

一个对难题更简单的解决办法来自今天宇宙是由暗能量主导的事实。如果这个暗能量真的是个宇宙学常数，并保持不变，那么随着宇宙继续膨胀，其中所有的物质及辐射都会被逐渐稀释，直至只剩下宇宙学常数。甚至由空间弯曲产生的能量都最终会消失——因此这样的宇宙就会变得平直。那么多年以后的宇宙学家是否会说根本就不存在平直问题，只是因为宇宙学常数为平直提供了一种机制吗？他们是否会认为宇宙平直问题其实是宇宙学常数问题，因为宇宙的平直性依赖于宇宙学常数的大小？

或者到那时宇宙中所有的恒星都已熄灭，根本没有宇宙学家会来问问题？

166

是否存在暴胀模型的替代品？

第十三章

大收缩和大反弹

在接近 t=0 的时候，你会问“大爆炸之前发生了什么？”或者“大爆炸之前是否有一次大收缩？”。实际上，也许你就是导言中那个在讲座后冲到讲台上提问的人，这个问题比“我们是不是位于宇宙的中心？”或者“宇宙要膨胀到哪里？”更加流行。

大爆炸之前发生了什么是个非常自然的问题，宇宙学家自从发现宇宙在膨胀就一直在思考。曾经提出过很多设想，但至今没有定论。膨胀和收缩阶段交替的宇宙被称为循环宇宙模型，或者“反弹”宇宙学，在过去 10 年间，人们再次认真审视它，用它来替代暴胀宇宙学模型。

167

循环宇宙模型极具吸引力，因为它使得我们免于苦苦思索一个在过去的特定时间里，忽然间就无中生有地产生一个宇宙的困境。数学上讲，这意味着我们不需要特别给出宇宙的初始条件，因为根本就没有开端。但想象一个无止无休在膨胀和收缩之间振荡的宇宙，也绝非易事。

困扰循环宇宙模型的问题一直都是所谓的大爆炸奇点问题。

我们无法再回避了。在弗里德曼宇宙中，在大爆炸瞬间，温度、压强、密度以及宇宙膨胀速度都是无穷大。这是我们所能理解的最为彻底的系统崩塌——远比一场瘟疫或者经济大萧条严重得多，因为二者终归能够结束。在大爆炸瞬间，全部广义相对 168 论方程都在一片火海中失效，我们就是不知道在此之前发生了什么，而且可能永远也无法知道。弗里德曼自己察觉到爱因斯坦的方程是允许一个振荡宇宙存在的，但没有关注奇点。在 20 世纪 30 年代早期，物理学家理查德·托尔曼创建了一个包含更多细节的循环宇宙模型，他认识到了由奇点引发的致命难题，但通过假设奇迹发生来解决，认为宇宙在大收缩后可以再次膨胀。

✷

几十年来，宇宙学家一直相信比弗里德曼的宇宙更不规则的宇宙可能避开奇点。毕竟弗里德曼宇宙中物质分布均匀，如果宇宙是闭合的，那么空间就是球形的。在一个收缩的宇宙里，就好像一个收缩的球，所有物质会从所有方向逐渐迫近奇点，最终全部视线内的物质会在同一时间全部挤压在一点，这里的密度变得无穷大。但我们可以想象一个不那么对称的宇宙——也许是雪茄的形状。在这样的宇宙中，某个方向上的物质塌缩 169 可能比另一个方向更快，结果就避免了奇点的形成。

不幸的是，事实并不是这样，所有沿着这个思路的尝试都以失败告终。奇点还在那里。从根上讲，失败源自引力是种吸引力的本性，不管是否规则，吸引都会使物质都会聚集到一点。20 世纪 50 年代到 70 年代，由阿迈勒·库马尔·帕尔乔杜里、罗杰·彭罗斯和斯蒂芬·霍金提出的强奇点理论，证明在非常宽泛合理的条件下，大爆炸奇点是不可避免的。

但是所有的理论都会做出假设，大爆炸奇点可以通过引入足够大的排斥力而消除。宇宙学常数——暗能量——让星系彼此加速远离，正好提供了某种能够避开奇点所必需的排斥力。主要的问题是：到底宇宙学常数有多大，才能在产生大反弹的同时还不影响天文学观测结果？它真的是个常数吗？

例如，假定我们宇宙目前的膨胀是由一次塌缩引起的。那么在塌缩阶段，宇宙微波背景辐射会被加热，就必须假定存在一个足够大的宇宙学常数，能够让宇宙在达到 10 亿开之前发生反弹，10 亿开这个标志性的温度必须出现在大塌缩前三分钟。170 然而，在反弹——我们的大爆炸——后，不会有原初核合成，除非所有的轻元素已经按今天宇宙中的丰度存在，否则它们永远不会被创造出来。而且这样大的宇宙学常数会使得宇宙膨胀得太快，星系无法形成。因此，简单地加入宇宙学常数来解决弗里德曼模型的奇点问题不是个可行的办法。

可行的诀窍是在宇宙开端引入某种类似宇宙学常数的东西——也许就像第 109 页所示的势能图一样——并且它能够在引发灾难前消失。有太多这方面的提议，特征和动机各自不同，我们不在本书深入讨论那些栩栩如生的细节了。一个有吸引力的选择是，假设宇宙在开始塌缩至普朗克尺度（10^{-33} 厘米，见

第十二章）之前反弹，也就是大塌缩前 10^{-43} 秒的普朗克时间。

我们知道普朗克长度和时间意味着物理学的终结。在更小

171 的尺度和更短的时间范围里，我们关于时空的通常概念可能全都失效，取而代之的可能就是量子引力，用来描述或者通过奇点。量子力学实际上可以产生需要的排斥力，但正如已经提到的，量子引力理论并不存在。但如果宇宙确实会在普朗克时标前发生反弹，那么根本就没必要引入量子力学。在那种情形下，我们可以只依赖经典物理，这倒是确实存在的理论。

*

过去 10 年间，一些反弹派宇宙学家拓展了这些规则。就像暴胀理论，他们引入一个新的场，其行为好像宇宙学常数，能够引起反弹，但在这种模型里，反弹发生在大爆炸之后 10^{-35} 秒。这是一段普朗克时标之前漫长的时间（对物理学家来说）；它甚至是在大统一时期之前，在这种情况下，所有的经典物理学都完全适用。

你可能会想这样的宇宙模型是否可以解决暴胀理论要消除的宇宙学难题。事实是，有些难题能够解决，但大部分仍是老样子。

172 为了理解这一切是如何发生的，首先要认识到我曾在第十一章中给出的暴胀解决宇宙平直问题的直接解释——宇宙只是在一眨眼的瞬间就膨胀了 27 个数量级，从而看起来是平直的——其实是个谎言（尽管这是宇宙学家经常会干的“坏事”）。

如果我们站在海边面朝大海远眺，地球看起来是平的，那是因为我们的视界只有几千米，相对于地球直径来说实在太小了。但如果我们站在高山之巅，山的高度可与地球半径相比，那么显而易见，地球表面是弯曲的。

因此，“平直”是相对的；你必须总是将视界距离与地球直径相比。如果视界远比地球半径小，那么地球看起来就是平直的。同样在第十一章里，我们看到在一个塌缩的宇宙中，视界总是比宇宙本身收缩得更快，所以越接近大爆炸，宇宙看起来就越平直。

反弹宇宙模型也是同样的。当我们在一个塌缩宇宙中接近大反弹时，宇宙看起来会越来越平直，因为我们只能看到越来越小的距离。正是这块小小的平直的时空成为大反弹后我们现在的宇宙。

173

视界问题也可以如法炮制一并解决。如果你想象处在黯淡过去中的宇宙，就在前一个周期刚刚开始要再塌缩时，宇宙中所有部分彼此都已建立了因果联系，因为它们全部位于视界之内。随着宇宙走向大收缩，视界收缩的速度要更快，视界内的这一小块时空就成为反弹后我们目前的宇宙，正如暴胀理论阐释的那样。因为这一部分时空内所有的粒子在反弹前都已经建立了联系，所以不再有视界问题。

现代反弹宇宙学模型有一个异常特性，如果宇宙收缩非常缓慢，这一切问题都可以得到解决，因为非常缓慢的收缩可以使得宇宙塌缩阶段不必恰是膨胀阶段的反像。在某些模型中，宇宙甚至不必收缩太多就可以解决问题。而且正像前面章节提到的，指数膨胀并不是能够产生微波背景尺度无关谱的唯一机制，某些缓慢收缩的模型也能给出一模一样的结果。

174

还有不要忘了，暴胀理论预言的原初引力波并没有被探测到，而原初引力波正是由量子引力时期的涨落产生的。因为在反弹宇宙学模型中并不需要到达量子引力时期，因此根本也不会产生原初引力波。那些难以驾驭的量子涨落以及多重宇宙也不曾出现过。

当前，反弹宇宙学是非常活跃的研究领域，但历史告诉我们，那些活跃的研究有可能眨眼之间就发现自己已经被遗弃了。所以，现在决定大反弹宇宙学模型是否能解决因暴胀理论引起的那些令人头疼的问题还为时尚早，只能说当下，它们似乎是有可能成功的有吸引力的暴胀模型替代理论。

175

我们又如何知道这些理论是正确的呢？

第十四章

为什么需要量子引力？

我们已经到达了大爆炸后 10^{-43} 秒处，是时候——如果时间有任何意义的话——创建量子引力理论了。假如反弹宇宙模型被证明无法避免奇点，那么宇宙学家就没有别的选择了。然而创建量子引力理论的主要动力还不是奇点本身，而是物理学家长达几个世纪的坚定信仰，自然界所有形式的力都应该被统一进高耸的物理大厦中——传奇的统一场论。

已有的任何观测结果都不违背广义相对论，因此广义相对论被认为是科学理论正确性的极致代表。但它只是个经典理论，没有考虑量子现象。现代量子理论已被以与广义相对论同样的精度证实——哪种理论更优的争论仍在继续——但它们又没有考虑引力。 176

理论学家深入骨髓的信仰让他们坚定地认为，这两种迥异的理论一定能够合并为一致的量子引力理论。然而经过近一个世纪的努力，二者的结合仍然无果。用最浅显的表述来说，困难在于广义相对论是研究非常大尺度的理论，而量子理论的研究对象的尺度非常小。这种鸿沟很难逾越，但正像物理学家约翰·惠勒曾说的，有关量子引力最难的问题是：问题到底是什么？

那么让我们问几个基本问题；但不要期待有答案。

首先，什么是量子现象？量子力学和广义相对论到底应该在什么尺度上能够统一？虽然“量子”一词早已成为流行词汇的一部分，汽车品牌商和量子治疗师一直在努力，量子真正的含义仍然模糊不明。在经典物理学中，系统的大多数性质——例如能量——可以是任何数值。而量子力学的基本规则是，这些系统量不能取任意数值，而是离散的，或者说是量子化的，就像现金只能是便士的整数倍一样。麦克斯·普朗克于1900年为了解释黑体辐射谱而开创量子力学，正像第五章讲述的那样，他的基本假设就是黑体发出的光是量子化的，光的能量只能是光频率乘以一个常数的整数倍，普朗克把这个新的自然常数记为 h。现在通称为普朗克常数的这个数值，确定了所有量子现象（效应）的尺度大小。

177

1905年，爱因斯坦提出光不仅在普朗克尺度下是量子化的，而是光实际上就是与被称为量子的能量包有关，表现得和微观粒子一样。当普朗克谈论黑体辐射的光时，他实际上讲的是光量子——光子。一个光子的能量就是普朗克常数 h 乘以它的频率。一群光子的一致行动就表现为光波，当我们研究光波时，不再关注个体量子的性质。光波可以由麦克斯韦经典电磁学描述。

178

如果普朗克常数出现在某个理论中，那么可以说这是个量子理论。如果一个理论不包含普朗克常数 h，那它就是经典理论。无论如何努力，你都不可能在广义相对论中找到 h。另一方面，作为一种经典的引力理论，在广义相对论方程组中的每一个方程里，你都能发现引力常数 G，那是决定引力强度的常数。[①]

量子力学第二个重要特征是著名的“波粒二象性”。就像光可以表现为粒子那样，粒子也可以表现得跟波一样。每种粒子都有与其相关的波动特性。特别是，每种粒子都有波长，它取决于粒子的质量和速度——当然还有 h。把这个波长想象为粒子的量子尺度，也就是粒子表现得像波一样的尺度大小。对于电子那样的亚原子粒子来说，波长很小，大概就是原子直径，在日常生活中完全不引起注意。但在原子体系里，例如现代电子 179 学，物质的波动特性就变得极为重要了。

*

有了这些概念，我们就能够理解广义相对论和量子力学应该在何处结合——确切地说就是前面章节介绍的普朗克质量和普朗克时间。你可能了解任何测量单位，无论是米制、英制还是什么其他虚构的单位系统，都是基于三个基本量：质量、长度和时间。问题是，什么才是这三个基本量的最合适选择?

① 见第 11 页脚注。

物理学家乔治·J. 斯托尼在19世纪时提出，基于自然性质的测量单位是更合适的，例如电子电荷、光速 c 以及引力常数 G，这样的选择远比存放在巴黎的一根小棒好得多（译者注：指国际米原器）。随后，普朗克也基于同样的想法提出采用基本常数 G、h 和 c 作为测量单位系统的基础，该系统今天被称为自然单位或普朗克单位。稍稍耐心一点，你就可以用 G、h 和 c 表 180 达出一个长度，大约是 10^{-33} 厘米；也能给出一个时间，大约是 10^{-43} 秒；以及一个质量，大约是 10^{-5} 克。[①]

显然，普朗克长度和时间要比你（或者大多数物理学家）能想到的任何事物都更小，无法想象的小，而普朗克质量相对于亚原子粒子来说，又是无法想象的大——大得能够被现代天平所测量。如果你用光速的平方 c^2 乘以普朗克质量，你就得到普朗克能量，它大约要比大型强子对撞机产生的能量高 10^{15} 倍，后者是目前地球上能量最高的粒子加速器。

这些诡异的数值到底意味着什么？基本常数是宇宙中最重要的数值，因为它们决定了所有自然现象。引力常数 G 决定了引力的强度，普朗克常数 h 则划出量子效应是否变得重要的界限。当光速 c 出现时意味着相对论效应变得显著——物质在以 181 接近光速运动。

你很可能知道黑洞是一种特殊的天体，其引力场强到光都无法逃逸；黑洞的大小由其质量和常数 G 以及 c 完全决定，无

① 普朗克质量 $m_p=\sqrt{hc/G}$；普朗克长度 $\ell_p=\sqrt{hG/c^3}$；普朗克时间 $t_p=\sqrt{hG/c^5}$。

需任何其他因子。黑洞的大小可以被视为引力效应开始变得极端重要的尺度。如果你要求一种粒子的质量，它的量子大小——波长——与其引力大小相等，那么就得到了它的普朗克质量。这个量子黑洞的大小就是普朗克长度，而光穿过它所需要的时间就是普朗克时间。

所以，普朗克尺度代表着引力效应和量子效应同等重要时的长度、时间和能量。在这样的尺度上我们既不能忽略引力也不能忽略量子力学，而是必须创建量子引力理论来描述宇宙。

*

为什么这样一种理论如此难产？根本上说是因为广义相对论和量子力学两种理论的基本假设实在是太不一样了。量子力学忽略引力而广义相对论忽略量子力学。换句话说，量子力学假定时空总是平直的，像狭义相对论那样。广义相对论假定时空可以弯曲，弯曲程度取决于其物质成分。182

这是个很严重的问题，导致了极端的技术困难。量子力学最初被提出时，与牛顿物理学一样是有关粒子的理论。同样，量子力学也不考虑狭义相对论。而量子力学与狭义相对论的嫁接是 20 世纪 20 年代由保罗·狄拉克完成的，成果名为相对论量子力学。

然而相对论量子力学本身也还是有关粒子的——特别是电子，它们被视为点粒子。按照定义，点粒子没有尺寸。这一点导致了大麻烦的产生，当两个被视为点粒子的电子彼此接触时，

它们之间的电力变成无穷大。[①] 类似的，当接近一个电子时，这
183 个点电子的场能量也会变得无穷大，而由于质能方程 $E=mc^2$，它的质量也变得无穷大，必然包括了场的能量。

为解决这些难题的努力引发了量子场论。特别是量子电动力学，成为解释电子如何与光子相互作用的理论。非常朴素的愿望是通过把物质平滑为场，我们就再不必让两个点电子距离太近而导致各种无穷大，也就是这些奇点自然会消失。

量子场论中有一点小小的含糊，其假定所有相互作用都可以用粒子交换来描述——电磁力确实是由于光子交换。这样的粒子被称为虚粒子。我们可以把它们视为第八章中讨论的真空涨落的一种表现。根据不确定原理，由于真空能量一直有涨落，从不会真正为 0，因此它可以同时产生粒子对，只要它们的寿命不长于不确定原理允许的范围；这也是虚粒子得名的原因。所以，一个点电子周围会环绕着虚粒子云，希望以此软化奇点问题。

希望破灭。事情变得更糟，到处都出现无穷大。于是名为
184 “重正化”的数学方法出现了，用来解决理论的无穷大问题。终于有了确定的答案——其与实验结果以令人惊异的精确度完美符合，以至于量子电动力学被称为有史以来被检验过的最精确理论。

---

① 两个点粒子之间电力的计算公式与引力定律相似（见第 11 页脚注），只是把公式中的质量换成电荷量，引力常数 G 换成另一个常数。当两个粒子间的距离趋近于 0 时，它们之间的力就变成无穷大。

起初没人理解为什么重正化能有用，甚至重正化的创立者之一，理查德·费曼都把它称为“骗术”。今天重正化过程已经有了更坚实的数学基础，但无论何种情况，重正化仍被认为是对成功的场理论至关重要的；如果一种理论不能被重正化而给出可信的答案，那么它就会被放弃。

不幸的是，在标准的对引力量子化的过程中仍存在无穷大，而且重正化也失败了，理论无法给出有意义的结果。

✷

这个终极难题引发了大量尝试解决的办法，目标都是创立全覆盖的量子引力理论。最简单的路径是假定引力可以用广义相对论进行经典描述，而对问题中的任何其他场，例如光，都用量子场论处理。物理学家把这种办法称为“半经典”，不过是对这种退化策略的礼貌性称谓罢了。但是，当问题涉及的引力场不是太强——例如在一个足够大的黑洞周围（黑洞越大，其力场越弱）——时，这种半经典的办法还是很有效的。确定的是，半经典办法给出了量子引力理论最著名的胜利：斯蒂芬·霍金正是沿着这个思路于1974年做出了他的伟大发现，黑洞并不完全是黑的，而是精确地以黑体辐射的方式向外辐射能量。

185

因为能量太弱，黑洞辐射还没有被直接观测到。一个太阳质量的黑洞温度大约是千万分之一开，黑洞越大，温度越低，可以想象黑洞的辐射有多弱。实际上，当霍金的计算显示黑洞辐射是精准的黑体辐射时，大多数物理学家立刻就接受了这个令人高兴的结果。

如果黑洞向外辐射能量，那么它们必然会损失质量。随着质量损失，黑洞的温度就会升高；温度升高使得能量辐射更快，186 也就是越快地损失质量。这种逸散效应让霍金能够预测，黑洞最终会以壮观的爆炸结束生命。但霍金的方法假定黑洞的引力场也就是它的质量不会减少。因此这样的预测当然会被认为有猜测的成分。实际上，黑洞蒸发的过程应该对黑洞施加一种反馈，减慢进一步的蒸发；至少霍金的一位同事声称，已经证明这样的反馈会在黑洞爆炸前很久就让蒸发停止了。

结果可能是错的，但这个例子说明了问题有多难解决，而我们距离一个完整的量子引力理论又有多遥远。显然，霍金的办法不能应用于普朗克时间。

*

那么，还可能是什么呢？能应用在普朗克时间的是什么呢？

针对这一问题最著名的理论就是弦理论，这并非本书要谈论的内容。弦理论试图作为一种统一场论，或者人们常说的 187 “万物之理”，不仅将电磁力和核力统一起来（像 GUTS 那样），而且也包括引力。弦理论是量子场论，但其基础元件不是点粒子，而是极小的弦，其长度大约是普朗克长度。历史总是重演，再一次用有限的弦来抹平点粒子，希望排除无穷大。弦既可以是两端开放地振动，也可以闭合成环。普通粒子就是弦振动产生的泛音（谐波），正像小提琴的琴弦（或者管风琴）产生泛音一样。

弦理论中的弦与普通弦的一个主要区别是，普通弦存在于我们的宇宙四维时空（一维时间三维空间）中，而在一种弦理论里，弦存在的时空是十维的（一维时间九维空间）。多出来的空间维度像绕着一根圆柱一样卷起，尺度与普朗克尺度相当。这样的空间维度足够小，所以我们根本觉察不到它们的存在。

188 弦理论有很多数学上的成功。最值得庆祝的是弦理论能够推导出著名的黑洞熵，由雅各布·贝肯施泰因提出，霍金则给出更精确的计算。（我不准备讨论黑洞熵，但这个结论非常有名而且与黑洞有温度的理念密切相关。）弦理论还预言了引力场的交换粒子——引力子，我将对其作简短的描述。

弦理论中出现了普朗克长度，立即告诉我们它应该是一种可以用来描述极早期宇宙的理论。实际上这是极端困难的；目前为止，弦理论与物理学的其他分支学科还没有什么联系。特别是，还没有哪个地面实验能给出任何对弦理论的肯定支持。而且十维时空的观念是基于粒子物理的超对称概念，其将物质粒子（例如质子）与作用粒子（例如光子）混合成更大的粒子群体。不仅没有物理实验支持超对称理论，而且来自大型强子对撞机的结果似乎已经排除了超对称最简单的版本。

189 更糟的是，本来超弦理论的吸引力来自只有一个版本是数学自洽的。然而如今，可能存在 10^{500} 个不同的超对称版本，远比被称为弦理论景观的可能性还多。弦理论景观会让你想起第十二章中介绍的多重宇宙。你有理由认为任何能够产生 10^{500} 个宇宙的理论就是什么都没说。这真是个严肃的问题。

✷

另一个名气比弦理论稍逊的解决量子引力的理论，称为环量子引力理论。它无意解释万物，而是把自己限定在引力量子化的范围内。与弦理论有些类似，环理论的基础元件是普朗克长度的环——但环引力的环是四维时空的。事实上，它们不应该被视为存在于四维时空，而是为四维时空提供了基础的构建元件。环引力计算也能给出贝肯施泰因 – 霍金的黑洞熵。

在环引力理论中，讨论比普朗克长度小的尺度以及比普朗克时间短的时间都是没有任何意义的；空间和时间本身都是量子化的。把时空想象为柔软的晶格可能会有些帮助，它的可弯曲结构是普朗克长度和普朗克时间尺度的。更近似的，环引力理论更像是通常所称的“量子泡沫”，这是在很久之前就提出的理论。

190

我还不曾强调量子力学与牛顿物理学的第三个重要的区别：它与不确定原理共进退。量子力学是一种概率理论。牛顿力学中，只要我们精确地知道一个粒子现在的位置和速度，我们就能预言它将来的准确位置；量子力学与牛顿力学完全不同，它只能告诉我们在某个位置某个时间，一个粒子出现的概率。

在普朗克时期，也许就没有像“1 厘米”或者“1 秒”这样的确切存在。量子泡沫就要求对普朗克时期结束时“结晶”成为我们的宇宙进行概率描述。

量子引力论如何避免奇点的产生呢？量子涨落产生的压力与宇宙学常数产生的斥力非常相似。如果强度足够，它能够让宇宙在普朗克时期反弹。精确的结果依赖于具体的模型，这些 191 模型数不胜数。环量子引力论宣称能够解决这个问题，但还没有任何量子引力理论能够解决宇宙学常数问题——为什么今天的宇宙学常数这么小。

有一点是近乎可以确定的：为了与我们熟悉的场论相似，亦即力是由粒子传递的，那么任何量子引力理论都应该预言传递引力的引力子的存在。弦理论确实如此。尽管已经探测到了引力波，但单独的引力子还没有，也许永远也不会被探测到。如果中微子与普通物质的相互作用都是如此罕见，可以在击中任何普通物质之前毫无障碍地穿过几光年厚的铅，那么引力子与普通物质发生相互作用的概率要比中微子低大约20个数量级，使得直接观测到引力子几乎是不可能的。

这就提出了如何能以实验的方式验证量子引力理论的问题。有些物理学家认为并不需要用实验的方式验证一个理论的方方面面。我们可以把虚粒子视为只存在于大脑或数学中，只是用 192 来帮助我们想象场理论是如何运作的，虽然这些虚粒子无法直接被探测到。重要的是，理论预言的现象是可以直接被探测到的，而实验现象就可以证实理论。

另一方面，如果不能预言任何可探测的现象，那么它就只是数学自洽的理论了。由于极早期宇宙的理论和模型越来越偏离经验的领域，有些物理学家认为，接受一个理论的传统标准——它具有可证伪性或能够被证明错误——不再站得住脚了。取而代之的是，我们应该乐于以“元标准”来接受一种理论，例如该理论正确的可能性（如果这种概率有意义的话）或者甚至是其艺术价值。说实在的，数学上的优美一直是理论产生以及被接纳的背后的动力，但基于这样难以理解的性质的理论，被证实和被证伪的概率几乎一样。

193 近几十年来，理论物理学的风格和社会学都发生了巨大的变化，问题不可避免地涌现出来：宇宙学家是不是在自寻烦恼？你可能还会想起那句意第绪语谚语“人类一思考，上帝就发笑”。

我们是不是已经进入了后经验科学时代？

194 后经验科学这个词本身是否矛盾？

第十五章

多重宇宙和玄学

你已经耐心地等待有关多重宇宙的问题很久了。我也一直在耐心地等待。毕竟，没有哪个宇宙学讲座能够在这个问题被提出前结束。至于答案嘛，没有哪一个会比美国的宇宙学教父詹姆斯·皮布尔斯在2020年哈佛大学的讲座里给出的更好。他相信多重宇宙吗？

不。

本书结束。

就多重宇宙而言，事情就将如此。原则上，媒体和公众总是被最极端的猜测吸引，而日常工作中的宇宙学家通常不会太关注。然而，多重宇宙的概念已经闪耀在聚光灯下十余年了，对这类事情的思索引发的激动心情，正是年轻人投身宇宙学研究的原因之一。第十二章和第十四章中我们都曾提到，暴胀模型和弦理论显然都要求多重宇宙的存在。

195

但是，准确地说，这个“九头怪”宇宙到底是什么呢？“准确”这个词，无论在问题还是答案里，根本不存在。某种程度上，多重宇宙只是个语义学问题。按照定义，“宇宙”的意思就是“所有事物”，那么就完全没有多重宇宙存在的必要。现代宇宙学中“多重宇宙”的典型含义其实是拥有完全不同性质的“亚宇宙”的合集。有些宇宙是平直的；大多数则是弯曲的。在某些宇宙里，自然基本常数值与我们测量到的相同或相近。而在另一些宇宙里，这些常数与我们的测量值可能有多个量级的差别。有些宇宙中存在星系，另一些里则没有。我们生活的宇

196 宙是其中之一。[①]

多重宇宙是“后经验”科学的缩影——看起来没有办法用导言中描述的传统科学方法来检验多重宇宙的概念是否正确。确实提出过一些建议，但没有被严肃对待从而真地采取行动。宇宙学家搜寻暗物质是因为的确有间接证据表明暗物质存在，但他们不会去搜寻多重宇宙，因为没有任何证据支持。在讲座之后的问答环节，皮布尔斯说明了这种状况。

为什么我们生活在当下这个特殊的宇宙中，这是对疑问的宽厚的表述。如果更具体一些，问题就变成：为什么我们观测到的宇宙大约已经100亿岁了？

这是基本的人择原理问题。罗伯特·迪克的答案最为有名：“宇宙必定有足够久的历史，才能产生除了氢以外的其他元素，

197 因为我们都知道，要成为物理学家就必须有碳元素。”换句话

① 还有另一类多重宇宙，与量子力学相关。量子力学不能预言测量的结果，只能给出某个结果的概率。有些物理学家相信每一次测量都会让宇宙分裂，因此所有的结果都会出现，但是出现在不同的宇宙中。这被称为“量子力学的多重宇宙解释”。

说，如果宇宙的年龄不是至少已经几十亿岁，那么此刻我们就不会去观测它。更宽泛的人择原理则是，宇宙必然是我们观测到的那样，只有如此才能允许生命存在。一个不能产生生命的宇宙，显然也不会产生观测者。根据人择原理，正是生命的存在从多重宇宙中选择了这个我们所处的特殊宇宙。

*

20 世纪 70 年代人择原理盛行之时，人们对它的反应是怀疑和蔑视。很多物理学家认为人择原理只是同义重复；显然，我们的宇宙是这样，因为它可以支持生命。迪克和皮布尔斯给出了一个类比，可能会让人择原理显得不那么浅薄。在一场大型的俄罗斯轮盘游戏中，一群宇宙学家手里随机拿着有子弹或没子弹的手枪。然后一个聪明的统计学家出现了，经过筋疲力尽的一番统计计算后发现，活下来的宇宙学家手里拿的是没子弹手枪的概率超级高。

你可能会嘲弄地回应“显然嘛！”。然而这种抗议本身，就是承认情况依赖于有意的事后分析。对人择原理的反对主要是因为它不能预言任何事，因此不符合一个物理学理论的最基本要求。大型的俄罗斯轮盘游戏多少能让事情更清楚一些；结果本就可以预测到的。必须承认，当与宇宙玩轮盘游戏时，没办法在游戏开始前知道一个给定宇宙是否装了子弹。

198

但是有关人择原理一个著名的传说是天文学家弗莱德·霍伊尔曾在 1953 年用它来预言太阳内部的核反应必然存在，因为这样才能产生足够的碳以支持生命存在。然而，在霍伊尔当时的文章中找不到他提及人择原理的地方，这个故事看来是后来虚构的。

对于美国地质学家托马斯·钱柏林来说，则完全是另一个故事。19 世纪，在物理学家和自然学家之间展开了一场有关地球年龄的大辩论。以达尔文为代表的学者认为需要漫长的岁月才能让物种充分演化，而以开尔文爵士为代表的物理学家则不相信太阳能够以任何已知的产能机制来维持那么久的时间。199 1899 年，钱柏林认为开尔文提出的论据只能证明太阳正通过某种未知的能源燃烧，这种能源来自原子内部。事实证明，达尔文派和钱柏林是正确的，物理学家错了。钱柏林提出的依据可能导致了太阳核反应的发现。

*

近几十年来，人择原理被用来解释我们宇宙的很多特性，虽然只是马后炮。大多与我们试图限定上帝指纹的尺度和宇宙学常数的目的相关。我们知道，宇宙微波背景辐射的涨落大约是十万分之一。如果这个值过大，那么宇宙中的物质早已塌缩为黑洞。如果这个值过小，那么宇宙的物质还没有来得及形成星系和恒星。无论哪种情况，都不会在这样的宇宙中出现能够观测宇宙的观测者。

同样的，宇宙学常数使得宇宙加速膨胀，阻碍了物质塌缩 200 形成星系。如果宇宙学常数在星系形成时期远大于宇宙的物质成分，那时可观测宇宙大约是现在宇宙的1/5大小，那么根本就不会有星系形成。星系形成时期宇宙的物质密度大约是现在宇宙的125倍，因此那时的宇宙学常数就不可能比今天的值高1个或2个数量级。

人择原理最大的反对之声就是源于它很少能给出一个误差小于一个数量级的答案。这是事实。另一方面，能对宇宙学常数的取值范围给出几倍或十倍左右的限定已经是个重要突破了，总比第十二章中根据量子力学计算给出的120个数量级的差异要好得多。

很多物理学家，甚至是人择原理的提出者，都把它视为绝望之作。当我们处在大量理论都变得充满猜测的时代，它们似乎是不可避免的；认为一个理论如果拥有复杂的方程组，它就一定有意义，这样的想法是错误的。我们还应牢记人择原理只是个原理，并不是自然界的定律。在物理学的历史上，曾有很多原理被用来引导我们的思维，结果产生成功的理论；其中一些被证明比另一些更有用。宇宙学原理就被证明是非常成功的，即使它显然并不完全正确。但如果验证美之原理呢？物理学中关于美的想法，总是被纳入数学对称中——拥有规则图样的对称——虽然将对称的概念融入粒子物理已经被证实是非常成功的，但有可能已经超越了它的实用性。第十四章已经提到，大型强子对撞机还没有发现任何超对称的证据。

201

著名的最小作用量原理被物理学家广为接受。原理源自最简单的想法，即两点之间最短距离是直线，例如光就会沿直线传播。这条原理说，可以通过最小化被称为作用的量（与系统的能量有关），得到一个理论的方程组。历史上，最小作用量原理引发了多次物理学革命，已经成为所有现代理论产生的通用路径。物理学家会假定某种作用并把它最小化，从而得到某个理论的方程组，而不是从实验中得到正确的方程。爱因斯坦直到通过某种操作导出场方程之前，都不认为他的广义相对论是完整的。[①] 量子引力学的理论也起始于假定一种作用。然而正如我们所知的，最小作用量原理有时也会给出错误的答案。如果我们仅是对某个全新的理论假定一种作用，我们又如何知道方程组给出的结果是正确的呢？特别是当我们无法用实验检验结果时。

202

✷

如果有一把标尺，一头是美之原理，另一头是最小作用量原理，那么人择原理在标尺上的位置可能更靠近美之原理。而且我曾讨论过“弱”人择原理，如果不是同义重复，那么它看起来也并非没有道理。就像迪克的原版问题那样，它只是问及宇宙的某些方面——年龄——为什么是我们所观测到的那样。它假定已知的自然定律就是那样。而“强”人择原理则认为，自然定律必须是它们现在的样子。尤其是自然基本常数，例如 G 和 h，必须是我们测量到的值；否则我们所知道的宇宙就不可

203

① 数学家大卫·希尔伯特在这场竞赛中，以5年的领先优势击败爱因斯坦。

能存在。例如，如果自然常数与它们的实际值偏差太大，恒星就不能形成，因此很可能也就没有生命存在。

物理学家在接受“强”人择原理时有过一段艰难时段，因为“强”人择原理正是大设计论点的回响——宇宙时钟般的运作一定暗示着钟表匠的存在。确实，人择原理的极端版本更是要求宇宙最终必须产生生命。物理学家总是反对这样的想法，因为他们总是对目的论施以重拳，所谓目的论就是认为事情之所以会发生，是因为它们就是为最终目标的实现而发生的。自从亚里士多德起，科学就是沿着与目的论相反的方向前进的。

✷

如果没有人择原理，那么目前还不清楚到底如何从多重宇宙或者是弦理论景观中选择出可靠的宇宙。事态的现状无疑是因为缺乏实验或者观测证据，来对理论学家的想象做出任何限制。即使我们足够幸运能进化为先进文明，能够在实验室中创建出多个宇宙，从而检验多重宇宙和人择原理是否正确，在此之前还需要一段漫长的时间。 204

很有可能我们永远无法完全理解在普朗克时间或者大爆炸之前到底发生了什么，除非更新的反弹宇宙允许我们能够窥视那个时期。如果我们的理论最终也不能给出可观测的平滑转变，我们可能真地不得不依赖数学自洽、模糊的概率以及优美性来对理论做出限制。

同样的，物理学家不太可能创建出万能理论。千万别把这个名词太当真。即使那些试图提出万能理论的物理学家，也不会声称该理论能解释人类为什么会恋爱。然而，即使在其更有限的目标——将四种自然力统一在一起方面，它的实用性也不

205

是很明显。在追求万能理论的路上已经取得了很多成果，但很多科学家认为原则上，这种努力已经误入歧途。

取得最大成功的理论是那些应用于有限领域的理论。不了解宇宙最早期瞬间发生了什么，并不影响计算行星轨道。也许科学的最伟大成就是能够说点什么，而不是什么都说到。毫无疑问，万能理论仍然是不完整的。十维时空的弦理论，即使排除质疑而被接受，仍会留下为什么要有十维时空的疑问。没有哪个能解释万物的理论可以解释自己。无论是特定的自然常数，还是假定宇宙如何开端，总有些东西是要人工输入的。大多数宇宙学家都会承认，他们研究宇宙学并不是要解决自然的终极之谜，而是接近它们。放松放松，不要焦虑：未来的宇宙学家会继续琢磨……

206

为什么不是一无所有?

致　谢

感谢斯蒂芬·伯恩和帕蒂·威泽的批判性阅读。感谢匿名评论者对本书提出的有益建议。当然，书中的任何错误或疏忽，那都是作者本人的责任。

索 引

（以英文首字母排序，页码为每页的边码）

拓展阅读

本书的信息主要源自科技论文和研讨会，它们并不是面向大众的。以下列出的图书和文章均由备受尊敬的物理学家撰写，用语通俗易懂，可能比本书的专业性稍高。

1. 关于广义相对论的实验基础：

Clifford Will and Nicolás Yunes, *Is Einstein Still Right?* (Oxford University Press, 2020).

2. 关于大爆炸核合成的经典图书：

Steven Weinberg, *The First Three Minutes: A Modern View of the Origin of the Universe* (Basic Books, 1977).

3. 关于现代宇宙学，涵盖一些猜想：

Martin Rees, *Before the Beginning: Our Universe and Others* (Helix Books, 1997).

4. 关于宇宙背景辐射观测较新的读物：

Lyman Page, *The Little Book on Cosmology* (Princeton University Press, 2020).

5. 关于宇宙膨胀理论的来由：

Alan H. Guth, *The Inflationary Universe: The Quest for a New Theory of Cosmic Origins* (Basic Books, 1999).

6. 关于宇宙膨胀（弦理论）：

Roger Penrose, *Fashion, Faith and Fantasy in the New Physics of the Universe* (Princeton University Press, 2016).

7. 关于弦理论的入门读物：

Steven S. Gubser, *The Little Book of String Theory* (Princeton University Press, 2010).

8. 关于量子引力研究（个人观点）：

Lee Smolin, *Three Roads to Quantum Gravity* (Basic Books, 2001).

9. 关于人择原理（几乎是关注这一问题的所有人都想了解的所有内容）：

John D. Barrow and Frank J. Tipler, *The Anthropic Cosmological Principle* (Oxford University Press, 1986).

10. 关于暴胀和反弹：

Anna Ijjas, Paul Steinhardt, and Abraham Loeb, "Pop Goes the Universe," *Scientific American*, January 2017.

Paul Steinhardt, "The Inflation Debate," *Scientific American*, April 2011.

11. 关于暗物质研究：

Joshua Sokol, "Elena Aprile's Drive to Find Dark Matter," *Quanta*, December 20, 2016.

Daniel Bauer, "Searching for Dark Matter," *American Scientist*,

September-October 2018.

12. 关于引力波和马赫原理：

Tony Rothman, "The Secret History of Gravitational Waves," *American Scientist*, March-April 2018.

Tony Rothman, "The Forgotten Mystery of Inertia," *American Scientist*, November-December 2017.

托尼·罗思曼其他作品目录

非科幻类

Physics Mastery

Sacred Mathematics: Japanese Temple Geometry（与深川秀俊合著）

Everything's Relative, and Other Fables from Science and Technology

Doubt and Certainty, with George Sudarshan

Instant Physics

A Physicist on Madison Avenue

Science à la Mode

Frontiers of Modern Physics

科幻类

Firebird

The Course of Fortune

Censored Tales

The World is Round

A LITTLE BOOK ABOUT

THE BIG BANG

Tony Rothman

To my professors and colleagues,

who taught me more than they know

CONTENTS

INTRODUCTION

WHY IS THERE SOMETHING RATHER THAN NOTHING?

This is a little book on the biggest subject conceivable—the big bang. It is not a book about a television show. It is a book about cosmology. Cosmology, as cosmologists think of it, is the study of the structure and evolution of the universe as a whole. Over the past century, it has increasingly come to mean the study of the early universe: investigation of the origin of galaxies, analysis of the lightest chemical elements, observation of the heat radiation pervading all space, and exploration of exotic phenomena we can't directly see—dark matter and dark energy. Generally, cosmologists concern themselves with our universe in the first eons, years, and even fractions of a second after its birth. Cosmology is precisely the theory of the universe's origin: the big bang.

Cosmology is occasionally called the place where physics and philosophy meet. That is to an extent true, and to an extent unavoidable. When we get down to it, all science is the asking of questions and the pursuit of answers to those questions. If we pursue the questions far enough, we inevitably run out of answers. Cosmology is uniquely prone to this difficulty. When a conversation arises about the big bang, the first question any non-cosmologist (which is most people) asks is "What came before the big bang?" This is a natural and legitimate question, but it presently has no answer and that state of affairs is likely to persist past the shelf life of this author.

Nevertheless, my plan is to pose the questions asked by laypersons, as well as others, and attempt to answer them in the simplest manner I can. Since this is a book meant primarily for people who are curious about science but lack scientific and mathematical backgrounds, my colleagues will find it equally lacking in rigor and completeness, but my aim is not to cover as much territory as possible; rather it is to uncover a little territory if possible.

To that end, I have tried to keep technical jargon to a minimum, and although there will be enough numbers to satisfy anyone, no equation in the text is more complicated than one for a straight line; anything else, I've relegated to the few footnotes. I also assume that readers can understand basic graphs and are willing to follow some fairly detailed arguments. On the other hand, I agree with one of the countless aphorisms Einstein never uttered, "You should make things as simple as possible, but not too simple." Over the years, I have become convinced that there really is a level below which certain things cannot be simplified; in cosmology this is largely because of its inherently mathematical nature. If I cannot explain the mathematics in terms of a comprehensible physical concept, I won't try.

Despite the lack of anything resembling real math in this book, one of its aims is to convince you that modern cosmology is an extraordinary edifice built on rock-solid foundations and that you should become a believer. To that end, each chapter generally builds on the previous. You should start the book at the beginning. If your only interest is the bottom line, you will grow impatient.

As I've said, cosmology does raise profound questions. In exploring the conceptual underpinnings of the modern big bang theory, my hope is not to shy away from such questions. As a mentor once advised, "If you ask a stupid question you may feel stupid. If you don't ask a stupid question, you remain stupid."

Inevitably, as the book progresses there will be more questions than answers. After all, in pondering the imponderable it is a short leap from "What came before the big bang?" to the ultimate conundrum: "Why is there something rather than nothing?" Given that people have been asking this question one way or another for millennia without consensus, it is not reasonable to expect to find the answer here. Indeed, if you put that question to any honest cosmologist, the only reply you will get is "I don't know." An easier question is, "Do those equations on the white board of the TV show mean anything?" The answer is yes. Personal experience suggests that cosmologists are underequipped to answer any questions regarding cosmetics.

*

Because this book is intended for general readers, I will make use of analogies rather than equations. A danger lurks here because sooner or later every analogy breaks down. Analogies, like theories, are models of reality, not reality itself. In the case of the big bang, cosmologists usually resort to balloons to explain certain properties of the expanding universe, but the real universe is not a balloon and the analogy is imperfect. When considering analogies, it is crucial to locate the differences between the analogy and the reality.

I have already used the word *theory* several times. Let

me emphasize that when a scientist uses this term, it carries a different meaning than in daily life. The radio often informs listeners that a prosecutor has a certain theory about a crime, while the defense attorney has a theory that the prosecutor is crazy. Usually, these are conjectures made entirely without evidence and the situation changes too frequently to make any sense of it.

By contrast, a physical theory is a highly interconnected web of ideas and predictions underpinned by mathematics and firmly supported by experimental and observational evidence. When cosmologists speak of the big bang theory, they are referring to just such a web of predictions and observations. The elements of the big bang theory have by now been under scrutiny for an entire century, and so many precision observations support the overall picture that some cosmologists feel that their discipline already resembles engineering more than it does basic research. Believe in modern cosmology.

*

Yet a fundamental difference between cosmology and most other sciences remains: There exists a single observable universe. The essence of most sciences is experimentation and replication. A drug manufacturer tests a vaccine by running clinical trials on many subjects. If the results cannot be reproduced by scientists worldwide, the vaccine is not regarded as reliable. Cosmologists, at least at present, are denied the opportunity to run experiments on multiple universes and thus they cannot say with complete certainty how the universe would look had things started off differently than they did.

Nevertheless, although cosmologists can't say everything, they can say far more than nothing. Having a single universe at our disposal only makes it difficult when considering the universe as a whole, when addressing ultimate questions. Short of that, cosmologists draw on data and observations collected by their close cousins, the astronomers. Astronomers have traditionally investigated the behavior of planets, stars, and galaxies through earthbound telescopes or telescopes in near-earth orbit. Yes, astronomers are landlubbers, or might as well be; no spacecraft or telescope has yet traveled anywhere near the distance to the next star, yet alone another galaxy, which means it is impossible to perform experiments on astronomical objects. For good reason astronomy is termed an observational science.

The basic assumption underlying all astronomy, however, is that the fundamental laws of physics are the same throughout the universe. Astrophysicists, also close cousins to cosmologists and astronomers, have applied these laws to decode the behavior of stars and galaxies. Since it is impractical to send a space probe to the distant reaches of the universe, at least within the lifespan of a civilization, we have instead relied on light and other messengers to bring information from the far universe to us. It is, in fact, one of the great triumphs of modern science that we have been able to learn so much about the cosmos without going anywhere, by making this assumption that the laws of nature as we know them apply everywhere. To what extent the known laws of physics apply to the universe as a whole remains an open question.

Cosmologists attempt to reconstruct the evolution of the universe using the same approach as astronomers and astrophysicists: with pen and paper or computer, we apply established physics in a mathematically consistent way to model the system we are studying and check whether the results agree with observation. The system may be a cluster of galaxies or the whole universe. If the predictions of our model agree with the observations, we go out for a beer. If the predictions don't agree, we search for mathematical mistakes. If we find none, we search for conceptual errors. If, finally, no one's model agrees with the observations, we add new phenomena. If the new phenomena improve the results, we ask our observational colleagues to begin a search.

One thing any scientist should hesitate to do is add exotic phenomena to the current model before having exhausted more pedestrian explanations. In thinking about the earliest instants after the big bang, hmm. . . .

✷

At this moment you may be wondering exactly where astronomy and astrophysics leave off and cosmology begins. There is no precise boundary, and typically a scientist working in one of these areas knows a fair bit about the others. The difference is mainly one of *scale.* As mentioned, astronomy and astrophysics are traditionally concerned with the behavior of stars, planets, and galaxies, more recently with entire clusters of galaxies and even the superclusters—clusters of clusters of galaxies. A cosmologist takes the biggest picture imaginable, which begins somewhere around the size of a supercluster and asks how all this came to resemble the universe we observe. Although the physics governing the behavior of galaxies is the same as for stars, this book will not be concerned with those, or with planets. It will barely touch on black holes, as fascinating as they are. From a cosmological perspective, these objects are so small as to be insignificant.

Cosmologists find it extremely helpful to keep in mind the various astronomical scales. Throughout the book I will use the standard astronomical practice of stating distances in terms of the time it takes light to travel those distances. You may know that it takes light about eight minutes to travel from the sun to the earth. Call it ten. We can thus say that earth lies at a distance of about ten light-minutes from the sun. Similarly, a light-year is simply the distance light travels in one year. Astronomers never convert light-years to miles or kilometers, and you shouldn't, either. Rather, you should just develop a feel for the different scales found in the universe:

Four light-years is the distance to the nearest star beyond the sun.

The diameter of our Milky Way galaxy is roughly 100,000 light-years.

The distance across a cluster of galaxies is millions of light-years.

The size of a supercluster of galaxies is hundreds of millions of light-years.

The size of the observable universe is about fourteen billion light-years.

*

That is the scale of cosmology, the scale with which this book is concerned.

Can you give me advice on eyeshadow and mascara? **No.**

GRAVITY, PUMPKINS, AND COSMOLOGY

COSMOLOGY IS the study of how gravity determines the evolution of the entire universe, so to understand cosmology requires understanding gravity.

Gravity is by far the weakest of the known natural forces. To a physicist, a force is nothing more than a push or a pull exerted on an object—no "dark side" enters the picture—and one of the main reasons that physicists call their field the most fundamental of all sciences is that, over the centuries, they have learned that only four fundamental forces exist in nature. One of these, termed the *strong nuclear force,* is easily the strongest natural force and holds the nuclei of atoms together. Any atomic nucleus consists of neutrons and protons, and the electrical repulsion among the positively charged protons would cause the nucleus to fly apart were it not for the strong force binding it together. The energy associated with the strong force is what is released in atomic explosions. The strong force, however, operates only within the atomic nucleus, which is extremely small, as cosmology goes.

The second fundamental force is the *weak nuclear force.* Billions of times weaker than the strong force, it governs certain forms of radioactive decay. Tritium, the extra-heavy version of hydrogen, is radioactive and decays into a form of helium; its rate of decay is determined by the weak force. But like the strong force, the weak force operates only within the atomic nucleus, which is insignificant on the scale of cosmology.

In daily life the most important forces are the electric and magnetic forces, which are actually two aspects of a single *electromagnetic force.* This force is responsible for all of chemistry and operates in any device requiring electrical currents, from toasters to smartphones to everything we take for granted today. The electromagnetic force is the basis of modern civilization. But to produce electric or magnetic forces requires electric charges. Because astronomical bodies, such as planets, are electrically uncharged they exert no electrical or magnetic forces on each other.

All objects do gravitationally attract one another. Gravity, though, is almost unimaginably weak—that the gravitational tug of the entire earth cannot budge a refrigerator magnet is a hint of how weak it is compared to the electromagnetic force. The way physicists tend to state it is that the gravitational attraction between two hydrogen nuclei, protons, is about thirty-six orders of magnitude smaller than the electrical repulsion between them. In designing consumer electronics, engineers pay no attention to gravity.

Yet, because nuclear forces operate only inside atomic nuclei and because astronomical bodies are electrically neutral, it is left to the weakest force in nature to determine the fate of the universe.

Our modern theory of gravitation is Albert Einstein's general theory of relativity, which is often called the most beautiful scientific theory. This is true.

On a superficial level, we might regard general relativity as merely a refinement of Newton's theory of gravity, devised by Isaac Newton nearly four hundred years ago. It consists of a single immortal equation that shows how the gravitational force between two objects depends on their masses and the distance separating them. We don't even need to write the equation down to understand its message: knowing just the masses of the objects and their separation allows us to determine exactly the gravitational force they exert on one another.[1]

Above I said a force in physics is simply a push or a pull. More precisely, a force causes an object to change its velocity—in other words, to accelerate. If a piano is speeding up or slowing down, a force is acting on it. If the piano is moving at a constant velocity, no force is acting on it.

According to Newton, if we know the forces on an object, we know its acceleration, and can then completely predict its future behavior. Thus, if we knew the masses and present separations of all the stars in the universe, we would know everything there is to know about the universe's future—and its past, as well. For this reason, the Newtonian universe is often compared to clockwork. For the most part, it is.

*

Newton's theory of gravity works so well in ordinary circumstances that for two centuries astronomers believed it completely

---

① For reference, Newton's law gives the gravitational force F between two masses, m_1 and m_2 as $F=Gm_1m_2/r^2$, where r is the distance between them and G is the *gravitational constant,* a number that must be measured in the laboratory and that determines the strength of the force.

explained the motions of the solar system. In the mid-nineteenth century the first hints appeared that this might not be so. Like all the planets, Mercury travels around the sun in an elliptical orbit. If Mercury and the sun constituted the entire solar system, the point of Mercury's closest approach to the sun, called its *perihelion,* would always remain at a fixed point in space. Astronomers observed instead that the perihelion was gradually shifting its position over time. Calculations indicated that the gravitational tug from the other planets in the solar system could account for most of this shift, but a tiny amount was stubbornly left over. Many theories were proposed to explain the anomaly, but the ghost in the machine remained a mystery for over half a century.

When Einstein began work on general relativity in the early twentieth century, apart from Mercury's perihelion shift there was no observational evidence that Newtonian gravity might be inadequate. There was, however, James Clerk Maxwell's theory of the electromagnetic field.

You should first realize that Newton's theory is one of *particles* and *forces.* Two pumpkins sit in a pumpkin patch. We can think of them as two particles exerting a gravitational force on each other across the patch. Likewise, we can idealize the earth and moon as particles exerting a gravitational attraction on each other across space. In neither case does Newton's theory explain how the force travels from one particle to the other. For this reason, Newtonian gravitation is often called an *action at a distance* theory, *action* being the word for force in Newton's day.

Equally important is that the gravitational force between the two objects is evidently transmitted *instantaneously;* if the sun disappeared, nothing would be left for the planets to orbit and they would fly off into space with no delay whatsoever.

*

Instead of a pumpkin patch, imagine that the pumpkins are floating in a pond. We immediately feel the picture has changed. The water in the pond is composed of an enormous number of molecules, but they are so tiny we forget about them and instead think of the water as having a certain density and pressure at each point. Density and pressure are "bulk" quantities, making no reference to individual particles. This is a signature characteristic of a *field.* The air in a room can be regarded as a field. So can the elastic surface of a trampoline. A swarm of bees in many respects resembles a field.

The field picture provides a natural mechanism for transmitting forces. If the pumpkins are bobbed up and down, they create small disturbances that propagate across the pond as water waves. These waves are local disturbances traveling through the water field at finite velocities. By contrast, in Newtonian gravity, one needs to imagine forces that are somehow transmitted across great voids, infinitely fast.

"Objection!" you cry, politely: the gravitational attraction between the earth and the moon does not involve waves. True. All analogies break down. When thinking about the permanent gravitational attraction between bodies, whether we imagine forces or fields doesn't much matter. Nevertheless, fields exist; if you have ever sprinkled iron filings onto a piece of paper above a magnet, you have perceived the shape of its magnetic field fairly directly. On the whole, the field picture is so powerful that essentially all modern theories of fundamental physics are field theories. Without the field concept it becomes virtually impossible to describe electromagnetic and gravitational waves.

To be sure, when Maxwell considered the laws governing electric and magnetic fields, he was able to show that these fields could propagate through the vacuum of space in the form of an electromagnetic wave traveling at 3×10^8 meters per second.[①] His

---

① Scientific notation is indispensable in physics and astronomy. To clarify for anyone unfamiliar with it, the exponent indicates the number of powers of ten, or how many zeros follow the one. Thus, 10 can be written as 10^1, 100 as 10^2, and 1,000 as 10^3. 3×10^8 is 300,000,000, which shows why we use scientific notation.

discovery, published in 1865, astounded Maxwell, because that number was almost the exact speed of light, which by then had already been accurately measured. The conclusion was "scarcely avoidable," he wrote, that *light itself* must be an electromagnetic wave traveling not infinitely fast but at the finite velocity of 3×10^8 meters per second. Maxwell's prediction, the greatest theoretical triumph of nineteenth-century physics, was confirmed several decades later by the discovery of radio waves.

At the opening of the twentieth century, a number of physicists attempted to create field theories of gravity based on Maxwell's electromagnetism. They all failed, because gravity doesn't behave exactly like electromagnetism. Einstein was the first to understand the difference and the first to get gravity right. To appreciate how his theory, which he called general relativity, describes the gravitational field, we must first get a feel for the theory he had developed earlier that serves as the point of departure for general relativity: the special theory of relativity.

What is relative and what isn't?

A SPECIAL THEORY

FROM THE 1820s onward, natural philosophers understood that electricity and magnetism are intimately related. Electrical currents produce magnetic fields and vice versa. With his theory of electromagnetism, Maxwell showed precisely how this took place. In creating his special theory of relativity, Einstein showed that electricity and magnetism were not only related but were two aspects of the same phenomenon. In doing so, he discovered that Newtonian physics must be modified.

But Einstein would never have agreed with the famous adage "everything's relative." At bottom, virtually all physics concerns motion and the essential question asked by relativity is: What changes when something's state of motion changes, and what stays the same? Some things change while others remain the same, and the theory of relativity might just as accurately have been called the "theory of absolutes," which was in fact proposed.

The main thing that is absolute in relativity is the speed of light. The strange thing about Maxwell's discovery that electromagnetic waves travel at 3×10^8 meters per second in a vacuum is that this number, nowadays universally designated by the letter *c*, merely popped out of his equations. When we measure the velocity of a train or a baseball, it is always with respect to some other object. If we were standing in a country field we might see a train moving east at one hundred kilometers per hour with respect to the ground. From a car, however, itself moving east on a road parallel to the track at seventy-five kilometers per hour, the train appears to be moving at only twenty-five kilometers an hour. The velocity we measure of any body always depends on our *frame of reference*—roughly speaking, our vantage point or, a little more concretely, the place where we are standing.

Maxwell's result was strange because it merely says that $c=3\times10^8$ meters per second. With respect to what? Maxwell himself assumed that his electromagnetic waves propagated through the *luminiferous ether.*

Water waves travel through water and sound waves travel through air, so it was natural to surmise that light waves must also travel through a medium. The luminiferous, or light-bearing, ether pervaded all space and provided an *absolute standard of rest*. If you are sitting in a train, you are at rest with respect to the train, but the train is moving relative to the earth and the earth is moving relative to the ether. Mercury also has a velocity with respect to the ether, and you can compare the earth's velocity to Mercury's by saying each has its own *absolute velocity* relative to the ether. Maxwell believed that the absolute velocity of light relative to the ether was 3×10^8 meters per second.

Unfortunately, simple calculations gave the mysterious ether rather strange properties. For instance, if the ether were one hundred times thinner than air, it must be one thousand times stiffer than diamond. More to the point, all attempts to detect it failed.

*

In 1905 Einstein took the bull by the horns and declared the ether null and void. Furthermore, he accepted Maxwell's result that the speed of light was a constant, *c;* let this be a law of nature. Thus was born Einstein's *special theory of relativity.* It is based on two simple postulates.

The first: Absolute motion does not exist. Einstein took over this axiom from Galileo and it says that no experiment done on a train can decide whether the train is at rest or moving at constant velocity. All motion is measured with respect to some frame of reference, and no reference frame is preferred over another.

The second: Any observer in any reference frame measures the speed of light in a vacuum to be $c=3\times10^8$ meters per second.

A few Talmudic comments are necessary here. The first postulate is known as *the principle of relativity.* (Einstein didn't initially call his theory *relativity*; that name accrued to it over the following years, and the *theory of absolutes* was indeed proposed.) The theory is termed *special* because it concerns motion at *constant velocity.* Einstein did not address accelerated motions and assumed that the reference frames above are themselves moving at constant velocity. Motion is indeed relative in relativity.

The second postulate, apparently simple, changed everything. The idea that anyone in any reference frame measures the *same* speed of light directly contradicts Newtonian physics. If light behaves like the train passing the highway, then its velocity should depend on the reference frame of the observer, as physicists call any person or thing making a measurement.

*

The postulate of the constancy of the speed of light also showed that space and time could no longer be thought of as separate, as they had been for centuries. It is fairly easy to see why. Imagine a clock that consists of a ball bouncing up and down in a squashed train, as illustrated on this page.

Boris on the train sees the ball merely bouncing straight up and down and can define one second to be the amount of time for the ball to make a round trip from floor to ceiling and back.

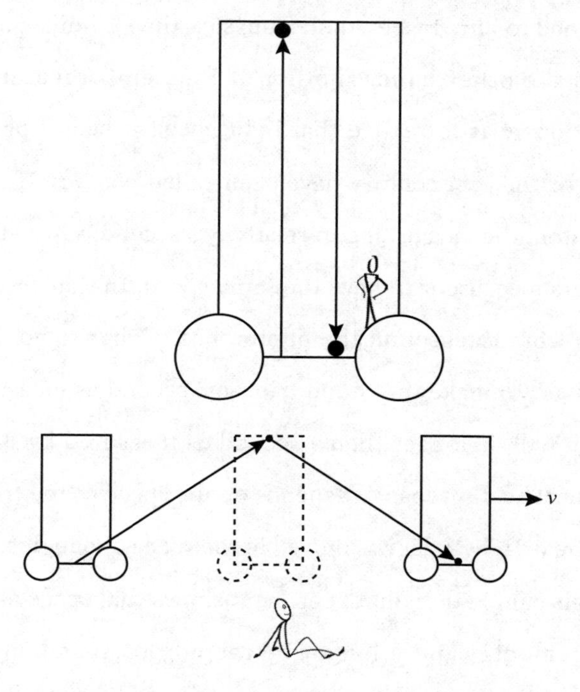

Natasha, however, observing the train from the ground as shown just above, sees the train moving to the right at its speed, v. One second to her remains the amount of time it takes the ball to make its round trip, but with respect to the ground the ball moves along a triangular trajectory, and thus travels farther.

But Natasha also sees the ball moving *faster.* It is bouncing vertically at the same speed Boris sees it, but to Natasha the ball is also moving forward at the train's velocity. Due to the extra speed, the ball covers the greater distance in exactly the same amount of time as measured by Boris, and one second to her is one second to him. In Newtonian physics, time is universal.

On the other hand, another of Einstein's revolutionary innovations was to realize that light is composed of particles, which for the past century have been called *photons.* If the ball is a photon, then according to relativity's second postulate, both observers measure it to have the same speed. In that case, since as seen from the ground the photon has farther to go, it must take longer to make the round trip. One second as measured by Natasha will be longer than a second as measured by Boris on the train. The discrepancy depends on the speed of the train and therefore on how far it has moved in the space of one tick.

This simple thought experiment shows that space and time measurements can no longer be thought of as independent. Einstein showed precisely how they are related, but for our purposes here those details aren't necessary. Since the advent of relativity, physicists no longer think of space and time separately;

instead they speak of a four-dimensional *spacetime,* which refers to combined distances in space and time.

Although the concept of spacetime is implicit in special relativity, Einstein was not its creator. Nowhere in his early papers on relativity did he refer to time as the fourth dimension. The French mathematician Henri Poincaré saw the necessity for spacetime earlier and the German mathematician Hermann Minkowski first worked out the implications. Einstein even opposed the idea as "superfluous erudition." Ultimately, though, the spacetime viewpoint proved essential to formulating general relativity.

*

Special relativity had other revolutionary consequences. One was that light provides the ultimate speed limit; no observer can measure a material object moving faster than light. Another is that as an object's velocity increases, its mass increases, to become infinite at c (which is one reason why nothing can travel faster than light).

Yet another consequence was Einstein's immortal $E=mc^2$, which says that the energy inherent in a body is equal to its mass times the speed of light squared. By definition, however, light travels one (1) light-year per year, so in that system of units $c=1$

and the equation says simply $E=m$. Since the advent of relativity, physicists have come to regard energy and mass as two aspects of the same thing, and they speak of "mass density" or "energy density" interchangeably, as I will.

Contrary to popular belief, Einstein was not the first person to show that mass and energy were related and, although it is heretical to say so, he never satisfactorily proved $E=mc^2$. His famous paper on the subject contains a mistake, which he attempted to patch up on subsequent occasions without success. Nevertheless, from its central role in explaining the operation of the atomic bomb or the nuclear reactions in the sun, the result has certainly withstood the test of time.

What has been left out of special relativity?

GENERAL RELATIVITY, THE BASIS OF COSMOLOGY

MODERN COSMOLOGY is essentially the application of Einstein's general relativity to the entire universe. By now general relativity has become one of the most precisely tested scientific theories, if not *the* most, in history. No experiment or observation has been made that contradicts it and there is no longer any question in cosmologists' minds that the theory provides an excellent description of our universe.

While the mathematics of general relativity is complicated, its basic concepts are accessible. Before turning to the cosmos, we should try to understand how a theory called general relativity became a theory of gravity, why we believe it, and how its viewpoint shapes our concepts of space and time.

If almost all physics is about motion, then in the past several pages we have overlooked something utterly fundamental: acceleration, the change in velocity. In creating special relativity, Einstein considered objects moving at constant velocity. Nothing accelerated, and since there cannot be acceleration without a force, no forces entered the picture, either.[1]

Einstein intended to enlarge special relativity to include accelerations—and in doing so, he created general relativity. If general relativity is often called the most beautiful theory (which is true), it is because despite the complicated equations, the entire edifice and all its predictions spring from exactly two simple yet profound assumptions.

Let us begin with what Einstein called the "luckiest thought of his life." Since Galileo's day it has been observed that when air resistance is negligible, all objects fall to the ground at the same rate. This is the famous acceleration of gravity, usually written *g*. Near the earth's surface, *g* happens to be 9.8 meters per second per second, but the numerical value is unimportant to those of us who are not engineers. To a physicist, the important thing is that *g* does *not* depend on the mass or composition of the falling

① With some work, accelerations and forces can be put into special relativity as it stands, but this does not transform it into general relativity.

object. Gold ingots, watermelons, and feathers all fall at exactly the same rate in a vacuum.

For this reason, if we were in an elevator and the cable were cut, we'd suddenly feel weightless, because we and the elevator are falling at the same acceleration, *g*, and our feet are no longer pressing against the floor, or on the bathroom scale we have conveniently brought along.

In a small confine, the state of free fall is indistinguishable from the absence of gravity.

This is precisely the situation in the International Space Station: astronauts and cosmonauts fall around the earth at the same rate as the station and thus feel weightless. A more common experience is that we feel heavier than normal when accelerated upward in an elevator. In this case, gravity seems to have increased.

Einstein raised these simple observations to the status of a law of nature, which he named the *principle of equivalence:*

In a small enough enclosure, no experiment can distinguish a constant acceleration from a uniform gravitational field.

In other words, if the elevator is windowless it becomes impossible to determine whether the cable is accelerating us upward, or the mass of the earth has suddenly increased and hence its gravitational field. ("Gravitational field" is another way of referring to the acceleration produced by gravity, *g*.) Likewise, if the elevator cable is severed, it becomes impossible for us to know whether we are really falling toward the earth with an acceleration of *g*, or the earth has disappeared altogether. Locally, accelerations and gravitational fields are equivalent.

For this reason, Einstein understood that to enlarge special relativity to include accelerations would require a new theory of gravitation.

*

Even more than the theory of special relativity, it was his theory of gravity, going under the misleading name of general relativity, that changed our notions of space and time. The principle of equivalence alone requires that clocks at different heights in the earth's gravitational field must tick at different rates. Not only does this demonstrably happen millions of times a day, but a good deal of modern life would be impossible if it didn't.

To slightly update a thought experiment proposed by Einstein himself, imagine a rocket ship accelerating upward in empty space. Natasha, at the top of the ship, would not be caught without a cell phone in her hand. At the bottom of the ship, Boris holds an identical model. Natasha's Equivalence App sends a light flash to Boris each second according to her phone's clock. But because Boris is accelerating upward during the transit time of the flashes, he is now moving faster than he was initially and intercepting them sooner than he would have, had he continued to move at a constant velocity. He sees the pulses spaced at shorter time intervals than Natasha does and therefore concludes that his clock is running faster than hers.① If accelerations and gravitational fields are equivalent, the same must take place in the gravitational field of the earth.

The Global Positioning System relies on timing signals provided by a constellation of satellites in orbit above the earth. Because the satellites are moving at high velocity, according to special relativity their onboard clocks are ticking more slowly than cell phone clocks on the ground. But because they are in high orbit where the gravitational field is weak, general relativity says they must be ticking faster than clocks on the ground. The discrepancy due to general relativity is actually twice as large as the one due to special relativity, but together they amount to something less than a billionth of a second each second.

At 3×10^8 meters per second, in one-billionth of a second light moves about a third of a meter, a foot. Unless the GPS corrected for relativistic discrepancies, each second your GPS

---

① Some readers may recognize that I am describing a Doppler shift.

position would get off by about one foot. Within a matter of minutes, those who no longer know how to read a map would be irretrievably lost.

General relativity is true.

*

It also provides a description of the cosmos Newton would never have recognized. You probably know Newton's famous law of inertia, taken from Galileo, which states that a body tends to keep doing whatever it has been doing. A little more precisely, if no forces are acting on an object, it travels along a straight line. Gravity causes objects to travel along curved trajectories, as when you toss a ball and it falls to the ground. But we have just seen that in a freely falling elevator gravity disappears. In that elevator, therefore, no forces are acting on a ball and according to inertia it must follow a straight path, as on the left of the figure on page 33.

Einstein decreed that light itself behaves in the same way. Thus, in a freely falling elevator, or one moving at a constant velocity, no forces are at work and light also travels along a straight line—again, as on the left of the figure. But in an elevator accelerating upward at g, or above a planet with a gravitational field of g, the equivalence principle requires that light *must* be deflected, and by the same amount in each case, as shown at the center and right of the figure.

How strange: it seems that whether an object follows a straight or curved path depends on the frame of reference, in the language of the previous chapter. Stranger still, it seems that whether gravity even exists depends on the frame of reference. This is true.

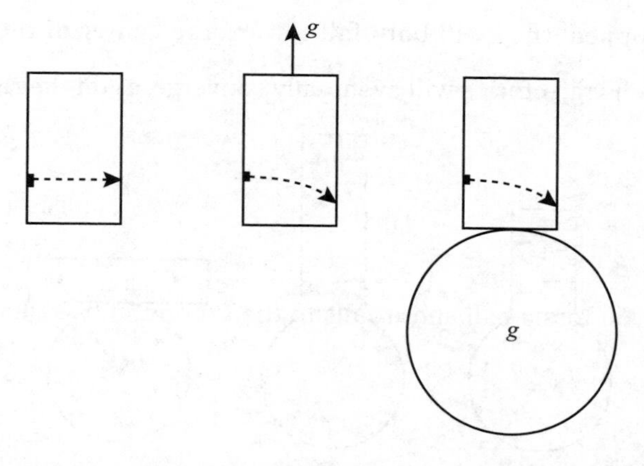

Imagine a building that may be built in the future, whose height is a substantial fraction of the size of the earth. At the top of such a structure, the earth's gravitational acceleration, g, is measurably smaller than at the bottom. This is no longer the "small confine" spoken of earlier.

If the cables are cut on two elevators, one near the building's top and the other near the bottom, they will fall at different accelerations. Someone who pitches a ball in the top elevator will see it move in a straight line, as will someone pitching a ball in the bottom elevator, but a person able to view both will see the balls following two different curves, which diverge. This is illustrated in the middle diagram above. Contrast this with the smaller building, on the left, where *g* is constant throughout and the two particles travel along identical trajectories, which never intersect. If the very tall building is lying on its side and two balls are dropped, they will both fall toward the center of the earth and their trajectories will eventually converge, as on the right.

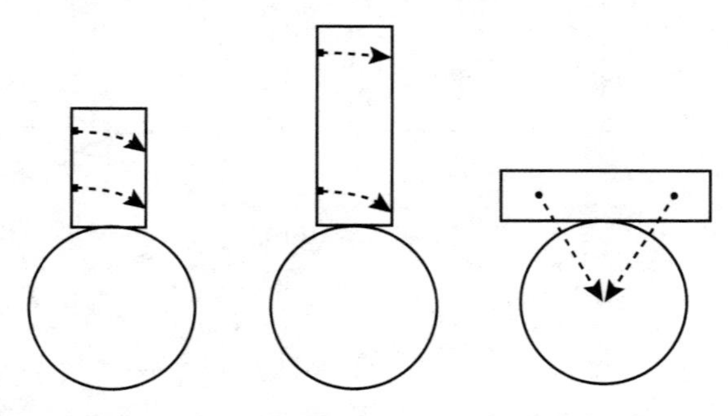

This situation in which nearby particles follow identical paths but widely separated particles follow different trajectories is one of *tides*. The side of the earth nearer the sun experiences a stronger gravitational field than the opposite side. The difference in forces results in a stretching of the earth, the famous tidal bulge, as well as ocean tides.

As we have seen, we can always find a small elevator in which gravity vanishes. Tides arise when we take a more global point of view, and as on earth, tides don't go away no matter how we look at the situation. In Newtonian language, tides are really the unambiguous manifestation of gravity.

Modern cosmologists describe gravity in geometric language. On a flat piece of paper, two lines drawn parallel to each other never intersect. Indeed, this is the famous fifth postulate of Euclidean geometry. In special relativity, no forces are at work anywhere and particles moving along parallel trajectories continue to do so forever. Special relativity is the theory of flat spacetime.

On a curved surface, however, two lines that are initially parallel may eventually intersect. Two lines of longitude are parallel at earth's equator, for example, but intersect at its north and south poles, as shown on the left side of the figure on this page. Notice also that the triangle drawn on the globe contains more than 180 degrees (since the base angles alone add up to 180 degrees). This is another sign of curvature. By contrast, two

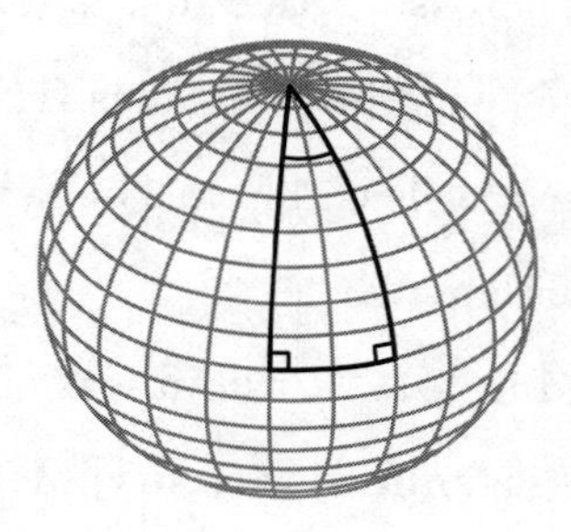

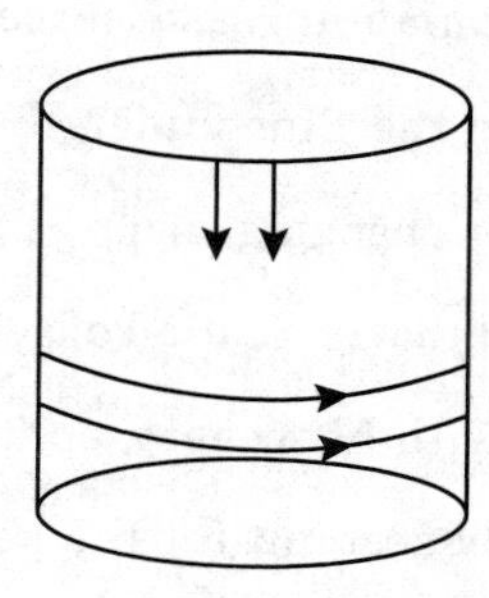

parallel lines drawn on a cylinder never intersect, and so the surface of a cylinder is not curved, despite appearances.

This is precisely the situation caused by gravity. Inside an elevator, particles follow parallel lines, but more widely spaced particles follow paths characteristic of curved surfaces, which may eventually intersect. Some physicists have regarded the geometric picture of relativity as an analogy irrelevant to doing physics. The geometry of general relativity, however, is exactly the geometry of curved surfaces, developed by Georg Bernhard Riemann and others in the nineteenth century, when that is extended to include time as a fourth dimension. If it is an analogy, it is a perfect analogy. Gravity *is* the curvature of space—that is, of spacetime.

Newtonian gravity tells us that massive objects produce gravitational forces and those forces cause other objects to move. General relativity tells us that matter curves spacetime and curvature determines how other matter moves. If in the Newtonian universe forces act across a space that is forever flat, in the Einsteinian universe spacetime becomes flexible, forever changing shape as matter travels within it. This was the conceptual revolution of general relativity.

With his theory, completed in 1915, Einstein was able to exactly account for the perihelion shift of Mercury; Mercury is the innermost planet and space there is curved enough to produce a measurable discrepancy with Newtonian gravity. In 1919 a famous eclipse expedition led by Arthur Eddington showed that starlight was deflected by the gravitational field of the sun, as Einstein had predicted. A century later, general relativity has become one of the most precisely tested theories in history. That map-reading has become a lost art is living proof.

*

General relativity, like electromagnetism, is a field theory and allows for the propagation of waves. As mentioned in Chapter 1, general relativity was not the first field theory of gravity and Einstein was not the first person to predict gravitational waves. In fact, he was initially a disbeliever, and even after coming around to their existence his first paper on the subject got it completely wrong. Nevertheless, he became the first person to get it right.

As in electromagnetism, where accelerating electrical charges produce electromagnetic waves—light or radio—in general relativity accelerating masses produce gravitational waves traveling at the speed of light. Gravitational waves are not light waves, however, and cannot be detected by ordinary telescopes. Rather, gravitational waves are tiny tidal disturbances propagating across spacetime, which stretch and shrink the measuring device itself, just as lunar tides do to earth. Because of the weakness of gravity, gravitational waves are unimaginably difficult to detect, stretching the detector an amount about ten thousand times less than the diameter of a proton. Nevertheless, after over a half-century of effort, researchers accomplished this miracle, and in 2016 the Laser Interferometer Gravitational Wave Observatory announced the discovery of gravitational waves. The wave patterns, caused by colliding black holes a billion light-years away, exactly conformed to general relativity's predictions and the discovery inaugurated a new epoch of astronomy, even as it caused tears to come to the eyes of certain cosmologists.

✷

Consequently, as far as anyone can tell, general relativity is as correct as scientific theories get. It is what physicists term a *classical* theory, meaning it takes no account of quantum mechanics. It may be necessary to create a quantum theory of gravity in order to describe the big bang *singularity,* a topic that will come up repeatedly soon enough. Barring that extreme event, however, general relativity works in every conceivable circumstance, and for that reason cosmologists do not hesitate to apply it to describe the evolution of the entire universe.

As we'll see, the real universe turns out to be nearly flat, or Euclidean, and therefore much of the formal apparatus of general relativity is nearly superfluous for modern cosmology; a Newtonian picture often suffices. Nevertheless, relativity's viewpoint is essential. In the vicinity of objects such as black holes, where the gravitational field can be extremely strong, spacetime is far from flat and there one must employ general relativity's full power.

*

Thus far I have said nothing about general relativity's second postulate. It has a rather inscrutable name, so let's just call it the "generalized" principle of relativity. Remember that special relativity concerned itself with motion at constant velocity—more precisely, with reference frames moving at constant velocity—and Einstein declared all such frames equally valid. None represented absolute space. In creating general relativity, Einstein declared that we should be able to describe motion in any reference frames whatsoever—in particular, in accelerating frames.

That declaration raises very deep questions.

Most of us have probably been on one of those amusement park rides that whirls us around in a rotating, circular cage, like a centrifuge. Indeed, we typically say that a *centrifugal force* has pushed us *out* against the cage wall. That's certainly how it feels. But a naysayer stationed on the ground would say, nay, it's a figment of our imagination. If the cage suddenly disappeared, we would fly off in a straight line as seen from the ground, in accordance with Newton's law of inertia. The centrifugal force we feel is "fictitious." In reality, the cage wall is pushing *in* on us, preventing us from flying off into space.

A spinning amusement park ride represents an *accelerating* reference frame and, according to many introductory textbooks, physics should not be practiced in such frames. The centrifugal force is fictitious because it disappears when the situation is viewed from the ground, which is not accelerating. Yet, we have already seen how gravity itself disappears in a falling elevator, which is equivalent to a nonaccelerating frame. Is gravity a fictitious force?

This question has an answer: If we believe in general relativity, we have no choice but to believe that either gravity is a fictitious force or that "fictitious forces" are real.

*

This raises an even deeper question. We sit in a train. According to special relativity, it is impossible to determine whether it is moving with a constant velocity or at rest, but we certainly know when it begins to accelerate—we are pushed squarely back into our seats.

With respect to *what* is the train accelerating? Isaac Newton would say with respect to absolute space—the ether, which forever remains at rest. Intro physics texts agree with Newton, and in doing so are saying that the ether really does exist.

In developing his general relativity theory, Einstein was strongly influenced by the German physicist and philosopher Ernst Mach, who believed that absolute space was a figment of Newton's imagination. Given that there is no way to detect absolute space, it only makes sense to talk about accelerations relative to other material objects—for example, the stars. Einstein christened this idea "Mach's principle."

The dilemma posed by Mach had already been famously demonstrated in 1851, in Paris, when Léon Foucault set a very long pendulum swinging from the dome of the Panthéon. As the day wore on, it seemed as if the direction of the pendulum's swings slowly rotated with respect to the Panthéon's floor. In fact, the Panthéon was rotating around the pendulum, which continued swinging in the same direction with respect to the stars above. How does Foucault's pendulum "know" to swing in a direction fixed relative to the stars? Or does the reference frame of the stars coincidentally happen to be the same as absolute space? Some people don't even see a question here. Others see one of the deepest mysteries of physics.

Einstein had intended to incorporate Mach's principle into general relativity. In a universe essentially devoid of matter, one would not be able to detect any accelerations at all. To what extent Einstein succeeded in this endeavor is debated to this day, but to explore it in any depth would require another book. So I leave it there.

How does relativity describe the entire universe?

THE EXPANDING UNIVERSE

TODAY, THE IDEA that the universe is expanding is so well known that it is part of our popular culture, but what does it mean? When audience members come up to the podium after any talk on cosmology, the first question is: "If all galaxies are moving away from us, are we at the center of the universe?" and the second question is: "What is the universe expanding into?" To be honest, sometimes these questions come in reverse order, but while they are natural, they show that the concept of an expanding universe is not.

It certainly was not to Einstein. When he published the general theory of relativity in 1916, there was no astronomical evidence that the universe was expanding, and when in the same year he applied the theory to create the first modern model of the cosmos, he assumed the universe must be static.

Over the next decade, astronomers were pushed to the idea of an expanding universe by the realization that nebulae—"clouds" often thought to reside within our galaxy—actually lay beyond the Milky Way; moreover, they appeared to be receding from us.

The acceptance of an expanding universe was clinched after 1929, when Edwin Hubble announced his famous "law" stating that the velocity of recession of a distant galaxy is directly proportional to its distance. For reasons that will hopefully become clear, Hubble's law implies that galaxies are receding not only from the Milky Way but from each other.[1]

This is exactly what astronomers mean when they speak of the expansion of the universe—galaxies are moving farther apart from one another. No discovery in cosmology has been more important and it lies at the foundation of the entire big bang theory. Surely, if the universe were not expanding, there could have been no big bang.

*

Conceptually, what Hubble did was simple: he merely plotted the velocities of a number of galaxies versus their distances. Despite his data resembling the points in the figure below, Hubble, being

---

① Recently, "Hubble's law" has been renamed the "Hubble-Lemâitre law," to include the Belgian priest Georges Lemâitre, who published it in 1927, but in French.

either very brave or very foolhardy, drew a straight line through them.

Here we must confront what is, I promise, the most difficult piece of mathematics in this book: the equation for a straight line. The equation for Hubble's line is $v=Hd$, where v is a galaxy's velocity, d is its distance, and H is the slope of the graph. The straight line implies that a galaxy's recessional velocity is directly proportional to its distance: If galaxy Beta is at twice the distance as galaxy Alpha, then Beta is receding from us at twice Alpha's velocity. Moreover, the greater the slope H, the faster galaxies at a given distance are receding.

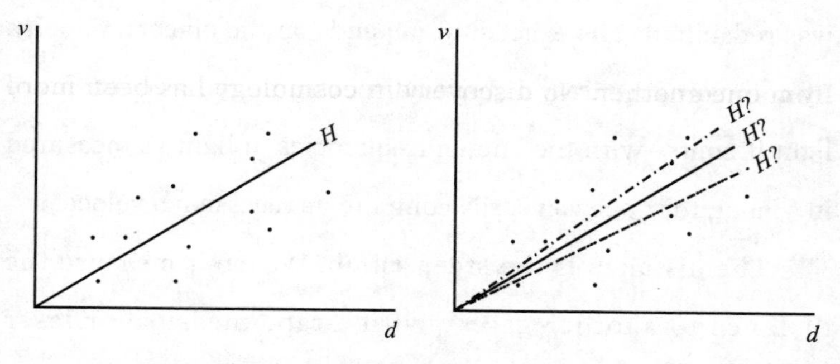

H, known as the *Hubble constant*, is easily the most famous number from cosmology and the careers of many cosmologists have been devoted to determining its exact value. Why is *H* so important? Knowing its precise value will not likely affect the outcome of elections, but in a way we'll see shortly, *H* measures how fast the universe is expanding, which enters into virtually every cosmological process. Furthermore, knowing H gives the age of the universe, the time elapsed since the big bang. In theory, to determine H is simple: following Hubble, plot galactic velocities versus their distances and read off the slope. The phrase "easier said than done" was invented for this task.

Measuring another galaxy's velocity is comparatively straightforward if we employ the famous *Doppler shift:* light frequencies from a moving object are shifted toward the red if it is moving away from us and toward the blue if it is moving toward us. Astronomers in the 1920s knew that most galaxies (or nebulae) were receding from us precisely because their light was redshifted. The exact shift depends on the object's velocity. By comparing a galaxy's observed spectrum—the frequencies of light it emits—with the known frequencies of light as measured in a laboratory, one can easily compute its recessional velocity.

The distance is the steep climb. We can't measure the distance to another galaxy with a tape measure or laser rangefinder. The distance to the nearest stars can be determined by triangulation, and the Hipparcos and Gaia satellites have extended this method to a billion stars of the Milky Way, but to measure extragalactic distances has required great ingenuity and sweat on the part of astronomers. The endeavor to establish the scale of the universe, the *cosmic distance ladder,* has probably been the major push of recent astronomy, but even with precision modern measurements, arguments over astronomical distances continue. As long as there are uncertainties in distance measurements, uncertainties will persist in almost every other astronomical quantity—in particular, in H.

That Hubble's own value for H was about seven times the modern number hints at the difficulties involved. Looking again at the figure on page 45, it is not altogether clear that the indicated slope on the left best fits the data; other possible slopes are shown on the right. For that matter, why draw a straight line in the first place?

✳

You can better understand the implications of Hubble's law by experimentally verifying it in your kitchen. Take a wide rubber band and mark on it a series of galaxies in the form of equally spaced dots. Label them A, B, C, D . . . . Stretch the rubber band until the dots are farther apart: A . . . B . . . C . . . D.

Pretend you are located on galaxy A. If the rubber band is stretching uniformly and B moves one centimeter from A, then C has moved one centimeter from B and hence two centimeters from A. Since this all happens over the time you have been stretching the band, C must be receding from A *twice as fast* as from B.

That is Hubble's law.

The key is that the band must stretch *uniformly,* at the same rate everywhere. Any universe that expands uniformly will exhibit a Hubble law.

I said above that H represents the expansion rate of the universe. Precisely, *H is the fractional rate of expansion of the universe.* In other words, *H represents the percentage increase in the distance to any galaxy per unit time.*

For example, if C is initially at five centimeters from A and it moves one centimeter in one second, then it has changed (1/5) of its distance per second and H is (1/5) per second. The rubber band can make this clearer, but I have put the demonstration in the note below.[①]

Most importantly, on the rubber-band universe no particular galaxy is any more central than another. If you were located on C, then A would appear to be receding twice as fast from you as B. The picture becomes even clearer if you imagine pasting galaxies onto the surface of a balloon. As you blow up the balloon, every

---

① Suppose galaxies A and C are separated by a distance d, and C has a recessional velocity v as measured by A. Since the rubber band obeys Hubble's law, $H=v/d$, *by* definition, velocity is the change in distance per unit time, usually written $\Delta d/\Delta t$. Thus $H=(\Delta d/d)/\Delta t$. This is the fractional change in distance per unit time.

galaxy moves away from every other galaxy, and all galaxies recede from their neighbors at the same rate. This is precisely what cosmologists mean when they speak of the universe's expansion.

So here is the answer to the first after-lecture question. Are we at the center of the universe? No.

You might well object that a balloon has a center—in its interior. Here is where the balloon analogy breaks down. A balloon is a two-dimensional surface in our three-dimensional space, and an ant on the surface can glance up into the surrounding room. The universe in which we live has three spatial dimensions and there is no surrounding room to look into. The real universe is a four-dimensional spacetime, not surrounded by anything else. The universe is growing larger, in the sense that galaxies are moving farther apart, but it is not expanding *into* anything. This is the answer to the second after-lecture question.

Of course, all this is terrifically difficult to visualize. In trying to imagine an expanding universe, people often picture in their mind's eye an expanding rubber sheet with an edge. Once we put an edge on it we are assuming an exterior, which does not exist. Once we put on an edge, we can locate a center, which also does not exist. Better is to imagine a sheet without an edge, stretching infinitely far into the distance. Galaxies marked on the sheet just keep getting farther apart from each other.

*

At this juncture you might ask: Are galaxies themselves expanding? Are you and I expanding? No, you and I are not expanding (except perhaps through dietary habits) because electromagnetic forces are holding our bodies together. Is the solar system expanding? The usual answer is no; the gravitational attraction of the sun holds the solar system together and prevents it from expanding with the universe. Similarly, galaxies themselves are bound by gravity and do not expand.

At larger scales, things become less clear, but at approximately the scale of superclusters, which can be a billion light-years across, the force of gravity becomes insufficient to bind objects together against the universe's expansion. Only parts of superclusters may be gravitationally bound, and the superclusters as a whole may participate in the expansion of the universe. The reason superclusters are the largest structures in the universe is because anything larger would not have formed a structure at all; the universe's expansion prevents it from coalescing.

*

Let's now run this entire chapter in reverse. If all galaxies are receding from each other, it is a fair presumption (though not a foregone conclusion) that at some moment in the past, this expansion began. The event that started off the universal expansion is what we call the *big bang,* a term coined derisively by astronomer Fred Hoyle in 1949.

The big bang was not a bang in the conventional sense; no one would have heard anything even had anyone been around to listen. It is also incorrect to imagine the big bang as a conventional explosion that took place in an already existing room. If there is no exterior to the universe, then there was no room for the universe to explode into. Spacetime as we know it came into existence at the big bang.

Finally, it is often said that at the instant of the big bang, all matter in the universe was concentrated at a single point, which must be the center. Because the universe does not have a center, this idea cannot be correct.

The rubber band can help sort this out. Assume that the band is already stretched and that A, B, C, and D are far apart. Relax the band until all the dots move back to their original position. The time for all the dots to return to their original position is the age of the universe since the big bang. Hubble's law tells us that the distance each galaxy crosses is $d=v/H$. But the distance a galaxy crosses is just its velocity multiplied by the travel time, $d=vt$, so $vt=v/H$, implying that $t=1/H$.

The inverse of the Hubble constant is known as the *Hubble age* and is the approximate time elapsed since the big bang.

Nothing here required all the dots to be at a single location. What's more, if we imagine the rubber band to be infinitely long, with an infinite number of dots A, B, C . . . (in an infinite number of alphabets), we are required to accept that the rubber-band big bang took place everywhere along this one-dimensional surface.

It is correct to say that at the instant of the big bang all the matter in the observable universe was concentrated at a single point. The observable universe, however, is not the entire universe. The distance that light has traveled since the big bang is termed the *cosmological horizon* and, as its name implies, we cannot see anything beyond it. We are permitted to say that at the instant of the big bang everything within the cosmological horizon was concentrated at a point.

Astronomers have devised many techniques for measuring the Hubble constant that are much more sophisticated than measuring the distances to galaxies. A few of these will appear in subsequent chapters. The difficulty is that these methods do not all agree. For now, let me say only that the age of the universe—the time since the big bang—is not quite 14 billion years, or to be overly precise, 13.7 billion.

*

General relativity's prescription to describe the entire cosmos, stripped to essentials, is this: determine the contents of the universe and how they are distributed; let the equations of general relativity tell you how the universe evolves.

That may be general relativity's prescription, but it was not Einstein's. As mentioned earlier, Einstein believed the universe to be static—nonexpanding. He forced his equations to produce such a universe by adding an extra term on mathematical grounds: the infamous *cosmological constant*. It was a pure fudge factor and , once the universe's expansion was established, Einstein discarded it as "the greatest blunder of his life."

In retrospect, adding the constant seems a strange move. If a fireworks rocket exploded in outer space, the cloud of particles would initially expand rapidly, and if the fireworks particles were massive enough, the expansion of the cloud would gradually slow due to the particles' mutual gravitational attraction. Depending on the particle mass, the cloud might eventually start contracting. One thing it would never do is stand still.

In the same way, applying the equations of general relativity to the cosmos without fudging shows that it is *dynamic*. A universe without any fudge factor will automatically expand or contract at a rate determined by the density of its contents. This indeed is the primary way in which general relativity reveals the effect of gravity—in determining the expansion rate of the universe. But just as Newtonian physics does not tell us how many fireworks to load into the rocket or what should be their composition, general relativity leaves open the ingredients for any proposed universe. Once they are specified, gravity takes over and guides the evolution of the model.

In 1922, Alexander Friedmann, a Russian meteorologist, produced just such a dynamic cosmos from Einstein's equations. Because Einstein was reluctant to accept an evolving universe, it is actually Friedmann's model that has provided the mathematical basis for the big bang theory.[1] The important feature of Friedmann's universe is that it is as simple as a cosmological model can get. It assumes that the universe's contents are uniformly distributed and the predicted expansion

---

① Friedmann's model was rediscovered over the years by Georges Lemaître (1927), Howard Robertson (1935), and Arthur Walker (1936), and so cosmologists today usually refer to it as the FLRW universe.

is uniform—that is, happening at the same rate everywhere.

Friedmann's main equation shows exactly how the expansion rate of the universe—the Hubble "constant"—depends on its contents. The Hubble constant measured by astronomers is actually *today's* cosmological expansion rate, which is technically only a constant at the instant you read this sentence. Generally, as the universe expands the density of its contents decreases, and with it, the expansion rate.

You may remember from Chapter 3 that the matter in the universe determines the geometry of space. If the density of matter in the universe exceeds a certain *critical value,* which is about 10^{-29} gram per cubic centimeter (say, ten hydrogen atoms per cubic meter), then, like the massive fireworks rocket, the expansion rate in Friedmann's model will slow to zero and eventually become negative—the universe will re-collapse. Such a universe is generally referred to as *closed,* and its spatial geometry is the geometry of a spherical balloon.

If the density is less than the critical value, the universe's geometry resembles an infinitely large potato chip (on which nearby parallel lines diverge) and it will expand forever. Such a model is generally termed *open.* As mentioned in Chapter 3, the real universe seems to be flat, exactly on the border between open and closed. With an expansion rate decreasing until finally becoming zero at infinity, the universe just barely creeps toward forever.[①]

If the expansion rate decreases toward the future, then it increases toward the past. Indeed, at the instant of the big bang it was infinite.

Surely that is impossible?

---

① In this discussion I am assuming that the cosmological constant is zero. If a cosmological constant is present, as is apparently the case in our universe (as will be discussed in Chapter 8), possible scenarios for the universe's behavior become more complicated. A spherical "closed" universe may expand forever and an "open" potato-chip universe may re-collapse.

COSMOLOGY'S ROSETTA STONE: THE COSMIC BACKGROUND RADIATION

IF THE DISCOVERY of the expansion of the universe was the foundation of modern cosmology, then the discovery that the entire cosmos is pervaded by a uniform bath of heat at three degrees above absolute zero laid the foundation of the modern big bang theory.

A few pages ago I claimed that the universe's expansion does not necessarily imply that the cosmos started in a big bang at some definite moment in the past. The universe might have always looked more or less as it does now—in which case, as galaxies recede from one another, new galaxies must be very slowly created to fill the voids. Such a scenario was once famous as the "steady-state cosmology," according to which the universe has existed forever.

While it is difficult to imagine a universe that has existed forever, it is equally difficult to imagine a universe popping out of nothing fourteen billion years ago. Until the mid-twentieth century, there was little observational evidence to favor either the big bang or the steady-state model.

That changed almost overnight in 1965. During the previous year, two radio astronomers at Bell Labs, Arno Penzias and Robert Wilson, had been employing an extremely sensitive antenna for the Echo satellite program to investigate radio emissions from our galaxy. For accurate measurements, one must minimize any local radio interference, be it from tractor spark plugs or the apparatus itself. To their mystification, after all conceivable sources of interference had been eliminated, including bird droppings in the antenna, Penzias and Wilson discovered that an unwanted signal remained. This weak signal appeared to be absolutely the same in every direction across the sky and so could not be from the galaxy itself. Penzias telephoned Robert Dicke, leader of the cosmology group at Princeton University, which had been readying a search for exactly this signal. After hearing him out, Dicke turned to his colleagues James Peebles and David Wilkinson and said, "Well boys, we've been scooped."

Penzias and Wilson had discovered the *cosmic microwave*

background radiation (CMBR), the very heat left over from the big bang. Remaining stalwarts of the steady-state model soon enough died off and the big bang theory became the standard cosmological model. The rest of this book will trace how the standard model has evolved.

✷

What exactly is the CMBR? All hot bodies, meaning all objects at any temperature above absolute zero, emit electromagnetic energy in the form of heat. Not only do ovens and computers radiate heat, but so do rocks, fish, you, and I. For historical reasons, physicists refer to pure heat as *black body radiation* and objects that radiate it as *black bodies,* even when they aren't black.

The fundamental and remarkable property of black body radiation is that nothing about it depends on the object's composition, only on its temperature. The body's temperature tells us the amount of emitted radiation and vice versa. When a doctor's assistant points a remote-sensing thermometer at your forehead as you enter a waiting room, what is being measured is the intensity of heat radiation you are emitting and consequently your temperature, assuming that you are a black body. Penzias and Wilson applied a remote-sensing thermometer to the universe and measured its temperature, which is now known to be nearly 2.7 degrees above absolute zero.

A typical FM radio station broadcasts at a frequency of about a hundred megahertz, corresponding to a wavelength of three meters.① Unlike a radio station, a hot body broadcasts radiation across all wavelengths, but the amount radiated at each differs greatly. For a black body, the intensity of energy emitted at each wavelength—its *spectrum*—is determined by its temperature and only by its temperature. For that reason, the black body spectrum is nearly universal. It resembles the graph above, although the exact shape depends on the temperature. As you can see, most of the radiation is given off near a peak wavelength, which for a 2.7 degree black body is just about .3 centimeters, or

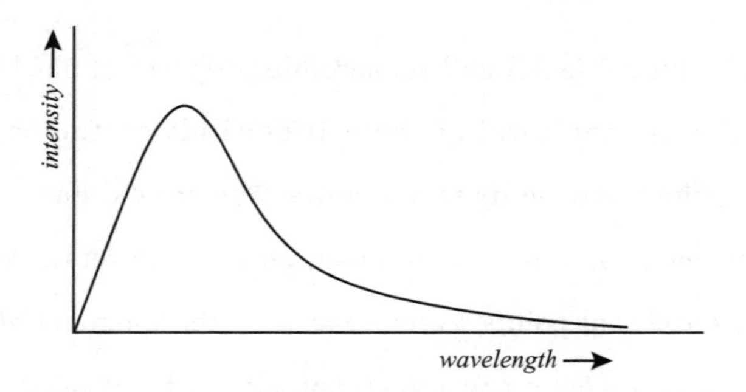

① One can speak interchangeably about frequency and wavelength. As frequency (f) goes up, wavelength (λ) goes down such that $f \times \lambda = c$, where c is the speed of the wave (3×10^8 meters per second for light, and about 340 meters per second for sound in air).

100 gigahertz. This is in the microwave radio band, which explains the M in CMBR.

Intensity of radiation is precisely defined as the amount of energy passing through an area of one square centimeter every second. Like the intensity of water emitted by a garden hose, it can also be thought of as the number of particles streaming through every square centimeter of space each second. Because heat is at bottom electromagnetic radiation—light—the particles in this case are photons. To say that the temperature of the CMBR is 2.7 degrees is equivalent to saying that in every cubic centimeter of intergalactic space there are currently about four hundred photons from the big bang.

Since its discovery, the CMBR spectrum has been measured by many experiments, beginning with the Cosmic Background Explorer (COBE) satellite launched in 1989, and it more perfectly matches a black body than any spectrum ever recorded in the history of civilization. In our twenty-first century, no one doubts that the CMBR represents the afterglow of the big bang.

*

The discovery of the CMBR sounded the death knell for the steady-state cosmology because it immediately implied that the universe was *hotter* in the past than it is today. The steady-state model, in which by definition the universe was always as it is observed to be now, simply had no straightforward way to explain the CMBR's existence.

The big bang was indeed very, very, very hot. Because the universe is expanding, the density of the matter and radiation within it decreases over time; conversely, in the past the density was higher. This includes photons, which in the past were squeezed much more tightly together than they are today.

Each photon was also more energetic. As the universe expands, the wavelength of light traveling from distant regions stretches along with it, and longer wavelengths translate into redder light. This is the famous *cosmological redshift,* also often referred to as the cosmological Doppler shift, as I did in Chapter 4. To say that light becomes redder with the universal expansion is equivalent to saying that the energy of the photons making up this light is decreasing. Conversely, in the past photons were more energetic than they are now. Since temperature is simply a way of measuring photon energy, in the past photons were at a higher temperature than today. When the observable universe was two times smaller than it is now, the temperature was twice as high. It's that simple.

*

These remarks have three important consequences. The density of ordinary matter in today's universe is roughly 10^{-30} gram per cubic centimeter, which amounts to about one hydrogen atom per cubic meter. By contrast, via $E=mc^2$, four hundred photons per cubic centimeter, each at three degrees, constitute a mass density of about 10^{-34} gram per cubic centimeter. That is ten thousand times smaller than the density of matter. Discounting other ingredients, a cosmologist would say that the universe is currently *matter dominated.*

This was not always true. Going backward in time, the density of matter particles and photons increase at the same rate, like marbles squeezed together in a contracting bucket. But each photon is becoming more energetic. Thus, when the universe was about ten thousand times hotter than it is today, at some thirty thousand degrees, the energy density of photons overtook the density of matter. Before that time, which would have been about 50,000 years BB (after the big bang), the universe was *radiation dominated,* meaning that its behavior was determined by the properties of photons, not matter. The situation is sketched on the next page. The distinction between a matter dominated universe and a radiation dominated universe will

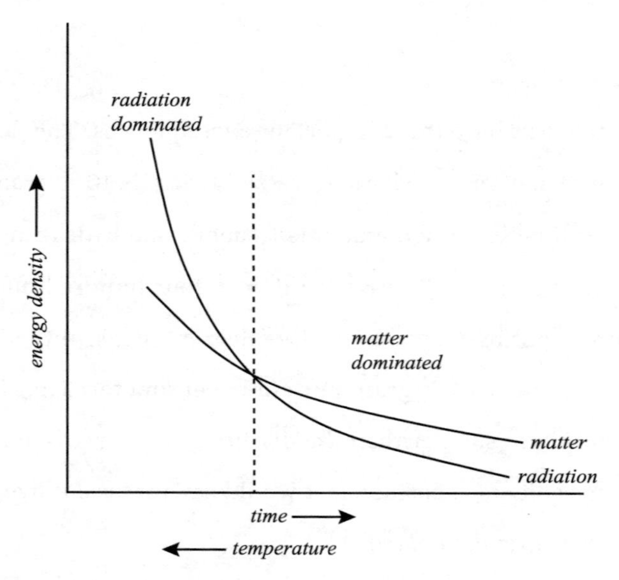

become very important, very soon.

A second important, not to mention disturbing, consequence of a hot early universe is that the temperature rise does not stop. Going back to one second after the big bang, the temperature would have been about ten billion degrees. At the instant of the big bang itself, the temperature would have been infinite. Infinities are rarely a good sign in physics. This particular infinity, like the infinite expansion rate that put an end to Chapter 4, is a manifestation of what is known as the big bang *singularity,* which will come up ever more frequently as we close in on the big bang. If the singularity at time-equals-zero really exists, it means that the theory has completely broken down. It's much like dividing by zero—illegal. We get infinity for an answer and the equations cannot predict anything further. Usually cosmologists start their thinking about the universe slightly after the singularity, when it was presumably behaving sensibly, if not comprehensibly.

*

Yet a third implication of a hot early universe is that the CMBR does not date from precisely the instant of the big bang.

About three-quarters of the mass of the visible universe is concentrated in the simplest element, atomic hydrogen, which consists of nothing more than an electron orbiting a proton. Because the electrons and protons carry equal and opposite electric charges, atomic hydrogen is neutral.

When the observable universe was at least a thousand times smaller than it is today, however, atomic hydrogen could not exist. The temperature was above several thousand degrees, high enough to "boil" electrons off their proton nuclei. More precisely, photons were energetic enough to knock electrons out of hydrogen atoms altogether, *ionizing* them. The resulting sea of detached electrons and protons is termed a *plasma*.

Photons cannot travel far in such a plasma because they immediately collide with the electrons and scatter. The result resembles what happens when you aim a flashlight beam through a dense fog: the beam is scattered in all directions, with the result that you cannot see far. In the early universe, as long as hydrogen was ionized, light was effectively trapped. As the temperature dropped to roughly three thousand degrees, the plasma cooled enough such that the electrons attached themselves to protons to form neutral hydrogen. Light does not interact much with neutral atoms, and after that time—which is strangely called *recombination,* although nothing was combined to begin with—light from the big bang streamed freely across the universe.

Thus, the CMBR as we observe it dates from the epoch of recombination, which modern measurements fix rather precisely at 380,000 years BB. Before this time, the universe was opaque to light, and by means of ordinary light we cannot see farther back in time than the birth of the cosmic microwave background radiation. Remember the term recombination.[①]

✷

When the CMBR was first discovered, its most important feature to cosmologists was its remarkable uniformity. Its temperature,

---

① The era of "recombination" is equally often called "decoupling," which emphasizes the cessation of collisions between photons and matter.

or the intensity of its radiation, as far as anyone could tell, was absolutely the same in every direction. Furthermore, on a large enough scale, galaxies themselves are distributed more or less evenly throughout the universe. Together these observations provided support for what has historically been known as the *cosmological principle:* on large enough scales, the cosmos is uniform.

The cosmological principle, buttressed by the featureless nature of the CMBR, became enshrined in the next iteration of the standard cosmological model: the model by which the universe started with a bang and this bang was absolutely uniform. No simpler picture could be imagined—but it had a number of great successes, the first of which will be discussed momentarily.

Such a simple picture could not be quite right, however, and today the most important feature of the CMBR is that it is not exactly uniform. In 1992, the COBE satellite observed slight irregularities in the temperature of the CMBR across the sky, which cosmologists knew must be there, or we wouldn't be here. These fluctuations represented the beginnings of galaxy formation. Perhaps you have seen maps of the bumps from COBE or its successors. The widely published sky map from the Planck satellite mission, launched in 2009, displays with unprecedented resolution the tiny variations in temperature of the CMBR. Although the irregularities represent a change in the background temperature of only about one hundred-thousandth of a degree, the size and distribution of the lumps has been key to unlocking almost every secret about the early universe.

What is so important about the cosmic microwave background radiation?

THE PRIMEVAL CAULDRON

CARBON, NITROGEN, oxygen, silicon, iron . . . these are elements we take for granted in daily life, and which are necessary for life itself. It is sobering to reflect that together such elements comprise far less than one percent of the visible mass of the universe. Most of the visible universe, some seventy-six percent, is made up of the lightest chemical element, hydrogen, and the second-lightest element, helium, makes up another twenty-four percent. Astronomy puts things in perspective.

One of the great achievements of twentieth-century astrophysics was the realization that stars are nuclear furnaces, transmuting hydrogen into heavier elements, including those above. Occasionally all these elements are scattered across space by supernovae, which in the process create even heavier ones, such as lead, gold, and uranium. Ultimately, the heavy elements are incorporated into infant solar systems, planets, and us.

Essentially our entire knowledge of the composition of stars comes from observations of their spectra. The spectrum of any light source usually contains distinct lines indicating the frequencies at which chemical elements within the source are emitting light. For instance, although most terrestrial helium is created from the decay of radioactive elements deep within the earth, helium is also observed in the spectra of stars. In fact, helium, from the Greek *Helios,* was first detected in 1868 in the spectrum of our sun. Modern observations of the earliest stars indicate they were formed with masses consisting of about 24 percent helium, as well as trace amounts of other light elements.

Because the earliest stars apparently came into existence with most of their helium already present, along with a few other light elements, the question arises: How were these elements created?

In the late 1940s, physicist George Gamow and his collaborators created what is now known as the *hot* big bang theory precisely to answer this question. And answer the question it eventually did. Its success in predicting the abundances of the light elements made it, after the discoveries of the universe's expansion and the CMBR, the third early triumph of the big bang picture and one of the pillars on which the entire edifice stands.

*

The theory of the formation of the light elements in the early universe, known as big bang nucleosynthesis, or slightly more poetically, primordial nucleosynthesis, is important not only because it gives results in good accord with observations, but also because in doing so it represents a successful fusion of general relativity and nuclear physics. It also gives the first answer to the question at the end of the last chapter, on why the CMBR is essential to cosmology. Indeed, even before it was discovered, the calculations of Gamow and his colleagues assumed this cosmic heat bath must exist.

The cauldron for primordial element formation is the Friedmann universe from Chapter 4, which is assumed to have uniformly distributed contents and which is expanding at a rate determined by those contents. In its broadest outlines, the entire element-formation process is simple: start with an expanding cauldron, add the necessary ingredients, cook.

A few pages ago I persuaded you that the past universe was hotter than today's. Indeed, for a few minutes after the big bang, the universe was hot enough to permit nuclear fusion reactions, which, not unlike those that take place in the sun, processed the available ingredients into helium. As the universe expanded, its temperature dropped and, "in less time than it takes to boil a potato," as Gamow once put it, the whole process ceased. The result was 24 percent helium and the observed amounts of the other light elements.

That's the high concept, but it is neither accurate nor complete, so let us delve into a few details, where the devil lies. The most important thing to remember is that there is nothing speculative about it; the entire scenario requires only conventional physics.

To keep myself honest, technically I have been speaking of light *isotopes*. Elements are designated by the number of protons they contain; isotopes of a particular element differ in their numbers of neutrons. An ordinary hydrogen nucleus consists of a single proton, whereas deuterium ("heavy hydrogen") is the hydrogen isotope consisting of one proton and one neutron. Ordinary helium consists of two protons and two neutrons and is called helium-4, while helium-3 consists of two protons but only one neutron.

Our goal is to produce in a very hot oven the astronomically observed abundances of these isotopes. First, the ingredients. To keep the recipe simple, we assume the material contents of the early universe to be exactly the basic building blocks that are found in the chemical elements today: neutrons, protons, and electrons. The cooking will be done by the four hundred photons per cubic centimeter (from Chapter 5) that comprise the CMBR.

There is one further ingredient: the subatomic particle called the *neutrino.* Neutrinos are the lightest of all fundamental particles, except for the photon, and they do not readily interact with other particles in nature. A single neutrino can pass through more than a light-year of lead before being stopped. For that reason, those neutrinos left over from the big bang have not yet been directly detected. One reason we know they must be present, however, is that without them the entire nucleosynthesis process could not have taken place, let alone give correct answers.

*

That is the entire ingredients list. Next, the oven temperature must be specified. To avoid pondering the big bang singularity, when the temperature was infinite, we pick a nonzero starting time. Let us imagine the universe at .0001 second after the big bang. Projecting today's CMBR temperature of 2.7 degrees backward, we find that, at .0001 second BB, the temperature of the universe was about one trillion degrees.

To talk about such numbers may seem fantastic, but in physics a lot can happen in a tenthousandth of a second, and a trillion degrees, while high, is not unimaginable. Ordinary protons and neutrons can exist at a trillion degrees and, moreover, the nuclear reactions among them are the ordinary ones known to physicists. At much higher temperatures, neutrons and protons would be "evaporated" into their constituents, the quarks, and no nuclear reactions could take place at all.

One trillion degrees, however, is indeed much too hot for atomic nuclei to exist. Protons and neutrons are colliding in this soup but moving too rapidly for the strong nuclear force from Chapter 1 to bind them into deuterium or helium nuclei. Just as temperatures above several thousand degrees ionize atomic hydrogen into a plasma of electrons and protons, at a trillion degrees atomic nuclei are "ionized" into a plasma of neutrons and protons.

But after about one second, the temperature has dropped to only ten billion degrees, which is roughly the temperature of the center of the sun and nearly cool enough for nuclei to begin sticking together. Assume for a moment that at one second BB there are seven protons for every neutron, as illustrated below.

At very nearly three minutes BB, the temperature has dropped to one billion degrees, which is frigid enough for colliding neutrons (n) and protons (p) to form deuterium (np).

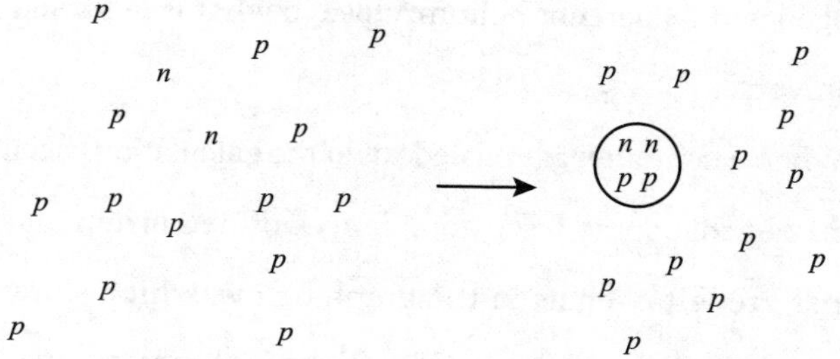

At that point, in a series of nuclear fusion reactions indeed like those found in the sun or in experimental fusion devices on earth, the deuterium is rapidly processed into helium-4, ordinary helium (ppnn).[①] Helium is an extremely stable element and the reactions essentially stop there. All of this takes a thousand seconds or so before everything has settled down—perhaps less time than to boil a potato, depending on the potato size.

How much helium is produced? If at the three-minute mark there are seven protons for every neutron, and all the neutrons are processed into helium, then the reactions cease once the available neutrons are exhausted. As you can see from the figure, the result is one helium nucleus for every twelve hydrogen nuclei (protons). But since a helium nucleus is four times as massive as a proton, this means that by mass we are left with 75 percent

---

① The main reactions are: $n+p \rightarrow d$; $d+d \rightarrow {}^{3}He+n$; $d+d \rightarrow t+p$; $t+d \rightarrow {}^{4}He+n$; ${}^{3}He+d \rightarrow {}^{4}He+p$; $d+d \rightarrow {}^{4}He$. Here *d* represents deuterium and *t* tritium ("extra heavy hydrogen"), which consists of a proton and two neutrons.

hydrogen and 25 percent helium, close to what is observed in the real universe.

When a computer is enlisted to do the calculations accurately, it turns out that some trace abundances of deuterium and other isotopes are left over, as in the graph below, which shows how the mass fractions of the various light isotopes evolve as the temperature of the universe drops. Getting only the helium abundance right would be a significant achievement, but remarkably, all the light-isotope abundances are in good accord with astronomical observations. This near miracle is one of the main reasons that cosmologists came to believe the big bang

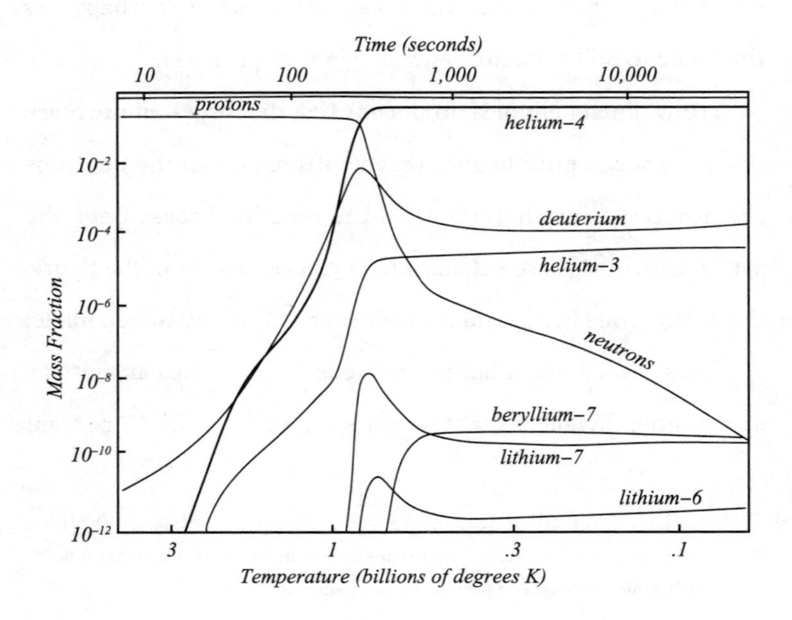

theory.

*

At this point, I hope you are asking where the peculiar ratio of one neutron per seven protons comes from. It is not too difficult to see how it comes about.

The first thing to realize is that neutrons and protons can be converted into one another. A neutron is essentially a proton plus an electron: $p+e\rightarrow n+\nu$, where the ν ("nu") represents a neutrino. The reaction can also proceed in reverse, converting a neutron into a proton: $n+\nu\rightarrow p+e$. Because these reactions are governed by the weak force mentioned in Chapter 1, they are referred to as *weak reactions,* and they show why neutrinos are an essential ingredient of nucleosynthesis.

In the early universe, because the weak reactions take place extremely rapidly, neutrons and protons are constantly being interconverted. At .0001 second BB, a proton is converted into a neutron faster than every billionth of a second. However, the neutron is slightly heavier than the proton, which means, according to $E=mc^2$, that more energy is required to create one. Consequently, there are always fewer neutrons than protons, but higher temperatures produce more neutrons.

Imagine a bunch of billiard balls bouncing around on a

billiard table, colliding with each other. The rate at which they collide depends on the number of balls, their size, and their speed, but on average there will be so many collisions per second. Now imagine that this billiard table is expanding. The bumpers are receding and so there are fewer ricochets. The table is stretching even as the balls move toward each other, resulting in fewer collisions. If the table is expanding fast enough, collisions will cease altogether.

Interesting things invariably happen in physics, and in life, when two scales cross. Large publicly funded projects may take decades to complete, but in the United States the federal government changes hands every four years; scales cross, projects are canceled, chaos ensues.

The early universe is much like an expanding billiard table. Its expansion rate depends entirely on the density of its ingredients. Projecting back from today's values shows that shortly after the big bang the density was by far dominated by photons and neutrinos. The density of neutrons and protons was so small by comparison that they played essentially no role in determining the expansion. In the words of the previous chapter, this was very much a radiation dominated universe.

At .0001 second BB, the neutron-proton interconversion governed by the weak reactions was taking place about a million times faster than the universe was expanding. As far as the weak reactions go, the universe might as well have not been expanding at all.

That situation quickly changed. As the temperature dropped, the weak reactions slowed extremely rapidly, and at about one second BB they fell below the expansion rate of the universe. Neutrinos stopped colliding with neutrons and protons and, as on the billiard table, the reactions ceased. The fraction 1 : 7 was the approximate ratio of neutrons to protons at this instant of "freeze-out"; thereafter, the number of neutrons didn't change too much before the onset of nucleosynthesis three minutes later.[①] The rest proceeded as already described, processing the neutrons and protons until the neutrons were exhausted at 24 percent helium.

Bear in mind that this entire discussion concerns only atomic nuclei. Atoms themselves did not form until recombination, 380,000 years later, when the temperature dropped to the point at which electrons attached to nuclei.

That the final abundance of helium is almost entirely determined by the neutron-to-proton ratio at "freeze-out"

---

① Free neutrons are radioactive particles and decay with a half-life of approximately ten minutes. Thus, about twenty percent would have decayed by the start of nucleosynthesis. The decay of the neutrons is charted in the graph on page 76.

enabled cosmologists in the 1980s to predict the number of neutrino types that exist in nature well before the number was established in the laboratory. That is, known neutrinos come in three species, called *flavors,* but the possibility of more flavors could not be ruled out. The existence of any additional flavors, however, would significantly increase the expansion rate of the universe during nucleosynthesis, which would in turn increase the helium abundance (because the expansion would overtake the weak reactions earlier, at higher temperatures, when more neutrons were present). Therefore, additional neutrino flavors imply a greater helium abundance. Limiting helium to the observed 24 percent ruled out new flavors, a prediction later verified in earthbound particle colliders.

Perhaps the most extraordinary thing about primordial nucleosynthesis, apart from the fact that it works, is that there are essentially no fudge factors. The conditions after .0001 second BB are within the realm of ordinary physics and the reactions are known from the laboratory. In the entire scenario only one number can be wiggled: the density of neutrons and protons in today's universe, which fixes their density at the time of nucleosynthesis. Since neutrons and protons are collectively known as *baryons* (for heavy particles), cosmologists speak of today's *baryon density.*

Now, stating the number of fatalities from a disease does not tell us as much as expressing it as a fraction of the population. In this case, the single input can be expressed as the ratio of photons to baryons. The photon-to-baryon ratio in our universe is roughly 10^9 to one, a billion photons for every baryon, and it is this number that produces such good results in the nucleosynthesis calculations. There is no understanding, however, of why this number is 10^9 rather than 1 or 618. Perhaps the universe merely started out with that photon-to-baryon ratio. Physicists, skeptics that they are, consider this a case of *fine-tuning*—in other words, adjusting the parameters of a model to make it fit reality. They prefer to find a natural mechanism to explain how the number arose.

"Naturally," one would expect the universe to have been created with equal amounts of matter and antimatter—there is no fundamental reason to prefer one over the other—but our universe is made almost entirely of what we term matter.[①] In 1967 physicist Andrei Sakharov suggested that during the big bang a slight imbalance of matter over antimatter arose—say,

---

① Antiprotons and antielectrons, for example, have the same mass as their matter counterparts, but opposite electrical charge.

a billion-and-one matter particles for every billion antimatter particles. *Star Trek* fans know that matter and antimatter annihilate on contact, producing two photons per annihilation. If a billion each of matter and antimatter particles were annihilated, one matter particle would be left over. We live in the "left over" universe, surrounded by a few billion photons per baryon. That explanation, however, only pushes the question back a notch: What determines the size of the matterantimatter imbalance?

Although Sakharov identified the necessary conditions for the imbalance to arise, a convincing explanation for the observed photon-to-baryon ratio has been elusive. This remains an unsolved problem of physics.

In general, we do not understand how the laws of physics arose. The very success of astrophysics is convincing validation of our assumption that laws concerning momentum, conservation of energy, and so forth are the same everywhere in the universe—and cosmology's success in describing processes like primordial nucleosynthesis is convincing evidence that the natural laws have not changed significantly since the big bang.

A fundamental theorem by mathematician Emmy Noether tells us that if a system is unchanging in time, then its energy remains constant—is conserved—and if space is completely uniform, then the system's momentum (*mass×velocity*) is also conserved. But this does not explain, for instance, how space came to be uniform, and does raise the question of whether we should enlist our usual physical laws (as we will in Chapter 11) to model the universe at extremely early times, before it became uniform. What's more, when we say that "energy can be neither created nor destroyed," we are invariably referring to closed, finite systems, like breadboxes. Bread can be turned into energy, and in doing so, its mass decreases, but what it means to talk about conservation of energy for the entire universe, especially if the universe is infinite, is not well understood, if it means anything at all.

Can we avoid fine-tuning the cosmos?

DARK UNIVERSE

MEMBERS OF THE PUBLIC rarely ask questions about primordial nucleosynthesis after lectures. Frequently, though, comes the query: "Can you tell me what dark matter is?"

The answer should be straightforward: No.

Let us end the chapter there.

Let us reconsider. Following the dictum that Einstein never uttered about making things "as simple as possible, but no simpler," a physicist's job is to cut through nature's red tape to create the simplest explanations of observed phenomena. But nature is rarely as simple as she first appears. As observations reveal increasingly complex phenomena, the models and theories required to explain them necessarily evolve from the simpleminded to the sophisticated. Nevertheless, in contrast to economists, physicists add complications with reluctance.

With the acceptance of the big bang in the years after 1965, the standard cosmological model became the Friedmann universe with its assumption of absolutely uniform contents. But COBE's discovery of ripples in the cosmic background radiation forced a revision of the standard model to account for galaxies, galaxy clusters, and superclusters, all of which undeniably exist.

Before attacking the new standard model in Chapter 9 and Chapter 10, we must first confront the existence of *dark matter* and *dark energy*, on which the model is partly based. Perilously, the situation regarding both changes by the week. In such circumstances it is wise to enlist the *New York Times* rule: if you read about a discovery in the *New York Times* before you have heard about it from a researcher in the field, don't believe it.

Communication satellites circle the earth only because gravity bends their trajectories into closed orbits, counteracting the satellites' natural propensity to obey the law of inertia and fly off along straight paths into deep space. Because the force of gravity on the satellite depends on the earth's mass, the satellite's orbital velocity does as well. The higher the satellite velocity, the greater the mass required to keep it in orbit. The same applies to planets in orbit around the sun or stars orbiting the galaxy's center.

The idea of unseen matter has popped up several times over the past century and a half. In the 1930s, astronomer Fritz Zwicky noticed that the velocities of entire galaxies in galaxy clusters were much too large to be explained by the luminous mass—meaning stars—within the cluster, and he proposed the existence of *dark matter* to make up the deficit. For the moment dark matter is, well, simply matter that emits no light. Zwicky's suggestion was not taken seriously until forty years later, when Vera Rubin noticed that the velocities of stars in orbit near the edges of galaxies were also too large to be explained by the luminous matter within the galaxies. The edge stars should fly off into intergalactic space.

The measurements made by Rubin and her team were straightforward. Employing the Doppler shift, it is easy to measure the velocities of stars circling the centers of their galaxies. By now such measurements have been performed on thousands of galaxies and clusters, and the results are invariably the same: most of the matter within galaxies is invisible. Indeed, about 85 percent of all the matter in the universe appears to be dark.

That much is nearly ironclad, and the afterlecture question is simple: What constitutes dark matter? The answer really is equally simple: We don't know. Anyone who says otherwise is either a salesman or a politician, not a scientist.

Anything that doesn't glow has been proposed as a dark matter candidate. There are so many contenders that this little book cannot discuss all of them—or indeed discuss any of them, because all candidates that have not been ruled out have not been found.

✷

Two natural thoughts for dark matter would be black holes, which by definition emit no light, and their cousins, neutron stars. Or perhaps "brown dwarfs," which are "failed stars" with masses of, say, several dozen times that of Jupiter. Brown dwarfs glow only faintly because they aren't massive enough to begin nuclear burning. Or perhaps Jupiter itself—many Jupiters—might constitute a portion of dark matter. Astronomers refer to such bodies collectively as MACHOs—Massive Astrophysical Compact Halo Objects. Unfortunately, MACHOS have been essentially ruled out as dark matter candidates, for good reason.

As discussed in Chapter 3, general relativity requires that massive bodies deflect light. Light passing around a star, a black hole, or a galaxy will be deflected from its original path, exactly as light is deflected by an ordinary lens. The result of such *gravitational lensing* is that the image of an astronomical object behind the mass-lens will shift its position or become distorted. By now gravitational lensing is a well-established phenomenon and many spectacular images have been taken by the Hubble and other modern telescopes.

Because the Milky Way is rotating, MACHOs near the edge of the galaxy rotate along with it. If light from some extragalactic source, such as an extremely bright star, passed near a MACHO acting as a gravitational lens, one would observe a slight twinkling of the star as the MACHO moved in front of it. Statistical studies of many stars in the Milky Way and Magellanic Clouds have not found any conclusive evidence for such gravitational lensing by MACHOs.

A more definite reason to exclude MACHOs is primordial nucleosynthesis. MACHOs, whatever they may be, are composed of ordinary baryonic matter (neutrons and protons), which was presumably present at nucleosynthesis times. Increasing the baryon density would increase the nuclear reaction rates forming helium during nucleosynthesis, leading to more helium. The abundance of helium actually observed by astronomers is produced when the baryon density corresponds to the *luminous* matter of the universe. If there is really five or six times more dark matter, it simply cannot reside in baryons; far too much helium would be produced during the big bang. This is a perfect example of how various aspects of a scientific theory reinforce one another.

Further, detailed analysis of the ripples in the CMBR radiation, coming in Chapter 10, requires the same ratio of dark matter to baryons as nucleosynthesis does. Whatever dark matter is, it is not the stuff we are made of.

*

That being the case, the next natural thought is neutrinos. Photons, light particles, transmit the electromagnetic force. Neutrinos are produced in situations involving the weak nuclear force and are not particles of light. They are light, however. For over half a century, in fact, physicists assumed that, like photons, neutrinos were absolutely without mass, which would of course rule them out as a dark matter candidate.

Beginning in 1998, however, that view began to change. Experiments in Japan's Super Kamiokande neutrino observatory revealed that the three neutrino flavors mentioned in Chapter 6 continually mutate into one another via oscillations. Such oscillations are analogous to the beats you hear when you hit a note on a piano and the strings are slightly out of tune. Just as the acoustic beat frequency is the difference between the frequencies of the individual notes, the rate of neutrino oscillations depends on the difference between the masses of the neutrino flavors. If the masses are zero, there are no oscillations.

Because neutrino oscillations do exist, we know that neutrinos have mass. Unfortunately, because neutrinos are such shy particles, putting an exact number on that mass has caused several decades' worth of headaches among experimental physicists. The oscillation experiments show a tiny mass difference, which suggests a similarly tiny mass, and experiments designed to detect the mass more directly indicate that a neutrino's mass must be at least half a million times smaller than an electron's mass, which is otherwise the smallest of any known particle. That in turn implies that the maximum mass for a neutrino is at least two billion times smaller than the proton or neutron mass. Measurements by the Planck satellite of the CMBR ripples suggest the neutrino mass must be smaller yet.

Consequently, in the most optimistic scenario the neutrino mass is incredibly small. But remember: there are about a billion photons for every baryon. And because neutrinos outnumber baryons by more or less the same amount (slightly less, actually) we know that, depending on the exact neutrino mass, the total mass in neutrinos could be a fraction of the mass in baryons. In the 2020s it is difficult to be certain about anything, but it seems unlikely that neutrinos can account for more than a small percentage of dark matter.

There are always "buts" in physics. In this case, there exists the possibility of a fourth species of neutrino, one that does not oscillate with the others and could have a larger mass. Such a neutrino goes by the name of *sterile*. But because the evidence for sterile neutrinos is currently inconclusive, I will leave them in peace.

*

For several decades the leading dark-matter candidate has been, in contrast to MACHOs, WIMPs, for Weakly Interacting Massive Particles. Like neutrinos, WIMPs do not interact by the electromagnetic force—in other words, they do not emit or absorb light—so it is possible they could be dark matter. They are assumed to be massive, somewhere between ten times and a thousand times the mass of a neutron or proton, and thus they can interact with ordinary matter by gravity or by direct collisions. A weakness of the proposition is that WIMPs are completely hypothetical.

WIMP searches have been ongoing for over twenty years. Typically, a WIMP detector consists of a cryogenically cooled tank of argon or xenon gas. A WIMP collides with a xenon atom, causing it to emit a minute flash of light, which is detected by sensors surrounding the tank. The main difficulties are two. First, a WIMP is not the only particle that can engage in collisions; cosmic rays or particles from the decay of nearby radio-active elements can do the same job, and such "false positives" must be excluded. Invariably, WIMP detectors are located deep underground, usually in old mines, to screen out the unwanted background. The second difficulty is that no one really has any idea of what they are looking for, which makes it challenging to design an experiment certain to snag the culprit.

Thus far, WIMP hunts have come up emptyhanded. In 2020 there was a moment of excitement when the XENON1T detector team in Italy thought it might have detected an *axion*.

Axions are regarded by many as the last best hope of dark matter. Named for a detergent, the axion was conceived in the 1970s by particle physicists to explain puzzling aspects of the strong nuclear force—specifically, why the neutron appears uniformly neutral even though its constituent particles, quarks, are charged. The axion is thought to be extremely light, even lighter than the neutrino, but in some scenarios of the early universe enough axions would be produced that they could constitute the required amount of dark matter. Such scenarios are themselves speculative, however, and since anything an author says about them is likely to end up being incorrect, let us leave it at that.

*

With so many negative results and so much conjecture, it would be surprising if scientists had not dreamed up alternative theories to compete with dark matter. To be sure, a handful of cosmologists reject the idea of dark matter altogether, suggesting instead that Newton's law of gravity be amended. At the edge of galaxies, gravity appears to be too weak to keep stars in orbit. Newton's law of gravity has not actually been tested at such large distances, however, so why not merely make it stronger out there? Such strategies are labeled MOND, for Modified Newtonian Dynamics.

One can indeed rewrite Newton's law of gravity to account for the behavior of stars at galactic edges, but this requires introducing a special length beyond which gravity is stronger than Newton would have it, and this length is equivalent to introducing a new constant of nature, analogous to the speed of light or the mass of the electron. Physicists make such moves with great reluctance. Furthermore, because Newton's law is the everyday limit of general relativity, any MOND theory requires modification of general relativity itself. Attempts to do this have been made, but at present it appears that all the attempts are inconsistent with observations. It is fair to say that most cosmologists regard MOND with far more skepticism than they do dark matter.

*

Having read this chapter, you may feel that it was less about cosmology than about particles. In a sense that is the point. The universe has proven to be an arena for exotic phenomena, and in our times one cannot divorce cosmology from the physics of elementary particles. General relativity, nuclear physics, elementary particle physics, and more have been woven together to create our present picture of the universe, and the various strands cannot be disentangled. It is well to understand that any new proposal in physics must contend with four hundred years of experiments and observations, and that inevitably nature turns out to be smarter than we are.

Have you forgotten dark energy?

DARKER UNIVERSE

NO, I HAVE NOT FORGOTTEN dark energy.

If it is sobering to realize that the matter of which we are composed represents only a small fraction of the matter that makes up the universe, it is even more sobering to realize that most of the universe may not be composed of matter at all. For the past twenty years, the majority of astronomers and cosmologists have accepted that most of the universe, by far, is composed of *dark energy.* The term is really nothing more than a place-holder; we have no idea what dark energy is, other than to say it is not matter and it accounts for about 70 percent of the energy content of the universe.

Perhaps this chapter should also end there, but to understand why most cosmologists believe dark energy exists, we must accept that Hubble's law, presented in Chapter 4, is a lie. That law, which states that the velocity-versus-distance graph for distant galaxies can be represented by a straight line, can be true only if the universe has been expanding at a constant rate for all time. In that case, Hubble's constant, *H,* is a genuine constant and Hubble's law holds: $v=Hd$.

On the other hand, one would naively expect that the gravitational attraction exerted by galaxies on one another should slow the universe's expansion. In that case, the most distant galaxies (whose light reaches us from early in the universe) should be receding *faster* than dictated by Hubble's law. The result is that the real graph should resemble what is shown on this page for a decelerating universe.

In 1998, the global cosmology community was shaken, to understate matters, when two research groups, the Supernova Cosmology Project and the High-Z Supernova Research Team, independently announced that, in fact, the universe's expansion

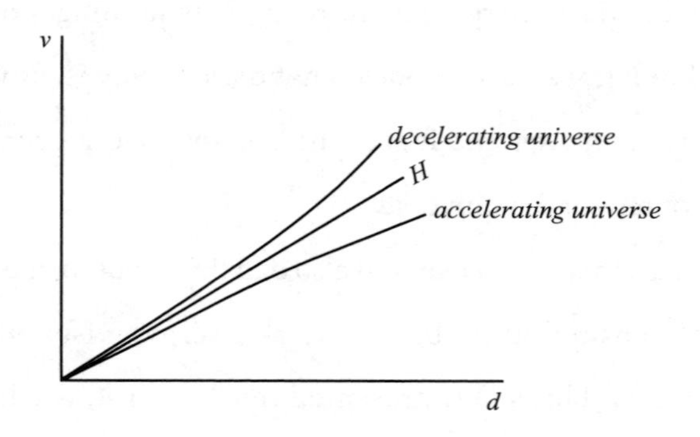

was not slowing down but speeding up. The universe was evidently accelerating. Cosmologists placed bets that the results would go away, like most unbelievable results in physics, but the rival teams' evident preference to die rather than collaborate lent credence to their results. Thus far they have stood the test of time.

What the researchers did was conceptually simple. Like Hubble, they plotted the velocity versus distance for many galaxies and searched for deviations from a straight line. As in the figure, such deviations do not show up nearby and so the teams needed to measure galactic distances across a fair fraction of the observable universe.

The key to making the notoriously difficult distance measurements credible was to find a *standard candle*. As we know from life, a light bulb appears dimmer the farther away it is. Specifically, the apparent brightness of a bulb decreases with the *square* of its distance from us: if the distance doubles, the brightness goes down four times, if the distance quadruples, the brightness goes down sixteen times, and so on.

If we observe two bulbs and see that one is four times dimmer than the other, we face a dilemma: we might be observing a twenty-five-watt and a hundred-watt bulb side by side, but we could equally well be observing two hundred-watt bulbs, one twice as far away as the other. If, however, we happen to know that the two bulbs have identical wattage, then one *must* be twice as far away as the other. Furthermore, if we know that each bulb is rated at a hundred watts, that tells us exactly how much energy it is putting out. Conversely, if we measure how much energy is actually reaching us—the apparent brightness—then we know how far away the bulb is.

A standard candle is simply a light bulb for which we know the rating. In the case of the supernova projects, the standard candle was a *type 1a supernova*. A type 1a supernova is produced when a white dwarf star that has been siphoning off matter from a nearby companion collapses, releasing an enormous amount of energy. In fact, such supernovae are billions of times more luminous than our sun and for a few days one can outshine all the other stars in its parent galaxy combined, making it visible across the universe.

A survey of many type 1a supernovae led astronomers to believe that, even if they did not exactly represent a standard candle, they could be adjusted to be one, with the result that when a Hubble graph was plotted, the universe's expansion appeared to be accelerating.

*

The acceleration implies the existence of some sort of force pushing galaxies apart. Frequently it is referred to as "anti-gravity," which is not helpful. Whatever the force is, it does not behave like gravity in reverse. For a time, the mysterious ingredient to the universe was often referred to as "quintessence"—Aristotle's fifth essence—which is an elegant term masking ignorance. More recently it has assumed the label *dark energy,* which does not explain much more, and should not be confused with the dark matter of the previous chapter. The two are not in any obvious way connected. One is matter and the other is, well, energy.

What dark energy closely resembles is Einstein's cosmological constant from Chapter 4, the fudge factor he inserted into his field equations to keep the universe static. Because Einstein used the Greek letter Λ (lambda) to signify his fudge factor, cosmologists today often refer to dark energy as the "lambda" term in the equations. Unlike gravity, the cosmological constant really is constant, and does not change as the universe expands. Contrary to the case of a static cosmology, in our universe Λ exerts an outward pressure that causes the expansion to accelerate.

We do not know how the cosmological constant arose. The general suspicion is that it represents the *vacuum energy* of spacetime, left over from the big bang itself. According to quantum mechanics, the vacuum of space is not empty but can be visualized as a roiling sea of energy. In physicists' minds, this sea of energy is pictured as a field of tiny, oscillating springs, which represent photons, neutrinos, and other particles. You have probably heard of the famous Heisenberg uncertainty principle, which is a law of nature. The uncertainty principle tells us that it is impossible to precisely know both the position and the velocity of a particle, or a spring, simultaneously. The energy of a spring depends on its stretch (position) and on its speed of oscillation. According to Heisenberg, these two cannot simultaneously be zero, so the vacuum springs always have some energy.

The difficulty is, if we estimate the total energy in these *zero-point oscillations* at the beginning of the universe, we find that it is at least 120 orders of magnitude larger than the dark energy today. Since that energy does not change, it remains 120 orders of magnitude larger than today's dark energy. This is the *cosmological constant problem.*

Cosmologists therefore face a choice: either Λ is not the result of quantum fluctuations, in which case no one has the slightest idea of how it arose, or one must devise a mechanism to drive it down to the value observed today, which is about fifteen times the visible matter density. Certainly if Λ were 10^{120} times larger than it is now, the universe as we know it could not exist. It would have been expanding far too rapidly for galaxies to form at all and primordial nucleosynthesis never would have taken place.

Consequently, if one believes that the cosmological constant was originally as large as simple estimates suggest, one must invent a mechanism to seriously decrease it, and very rapidly. Efforts to do so are ongoing, but there is yet no established solution.

A third choice, as usual, does exist. Recently, some cosmologists have disputed that type 1a supernovae can be used as a standard candle, suggesting that the observations are incorrect and that dark energy does not exist. That would be an elegant solution to the conundrum (if reminiscent of the flurry of excitement, in 2011, when the discovery of faster-than-light neutrinos was announced, only to have it turn out to be due to a loose connection in the equipment). Some cosmologists have other reasons for doubting dark energy, as well, but for the moment such voices are in the minority. In my hope to give this book a shelf-life longer than the time required for the ink to dry,

I shall not join the debate.

Actually, there is at least one further option. If the cosmological constant were so large that galaxies could not form, then almost certainly life could not exist in that universe. The very fact that we are here asking the question argues for a small cosmological constant. This is an example of *anthropic reasoning,* to which we will return in Chapter 15.

*

You may have noticed that the cosmological constant problem is similar to the question raised by the mysterious photon-to-baryon ratio of one billion to one, which cropped up in Chapter 6. Both problems ask for an explanation of the size of a number which has no obvious reason to be what it is. You may also feel that this sort of puzzle is of a different nature than, say, attempting to determine the value of the Hubble constant, which is a purely observational issue.

That is true. The photon-to-baryon ratio and cosmological constant problems are much more *why* conundrums than *how* problems. Traditionally it has been said that science is the province of how, not why, but over the course of the past century, as the gap between observation and theory has widened, the style of theoretical physics has shifted toward why.

Such questions invariably concern what physicists term *dimensionless numbers.* As briefly noted in Chapter 6, it is always best to express quantities in ratios. To claim that a certain presidential candidate won the election by 9,870,325 votes is almost meaningless. It becomes meaningful when you discover that 9,870,325 votes is 87 percent of the ballots cast, and then you might want to challenge the outcome. A dimensionless number is merely a ratio in which the units—dimensions to physicists—have canceled out, leaving a "pure" number. The density of lead is about 11 grams per cubic centimeter, or .4 pound per cubic inch. These numbers look very different from each other and do not tell us much. On the other hand, the density of lead—in the English system, the metric system, or the potrzebie system—is about eleven times the density of water. That is a dimensionless number. Now we are comparing apples to apples, or blintzes to blintzes.

The photon-to-baryon ratio of one billion to one, and a cosmological constant 120 orders of magnitude greater than the dark energy content of the universe, are dimensionless numbers. To describe the electrostatic force between two protons as 10^{36} times larger than the gravitational force between two protons is to use a dimensionless number.

To ask *why* these numbers are as large as they are is to

invite the response "because that's the way things are." One should not dismiss that reaction out of hand. On the other hand, physicists have it in their minds that all dimensionless numbers should "naturally" be about the same size, preferably near 1. If a particular number is orders of magnitude larger or smaller than all the others, that becomes an example of fine-tuning the universe to be what we observe it to be. Better is to find a reason that dimensionless numbers are the size they are.

In the history of physics, *why* has often enough become *how*. That many cosmologists regard the cosmological constant conundrum as the "most important problem in cosmology" shows that they take such matters seriously.

Are fine-tuning problems real or philosophical?

GALAXIES EXIST AND SO DO WE

OTHER QUESTIONS DEMAND immediate attention. The cosmological principle described in Chapter 5 insists that the universe should be uniform when viewed on large enough scales. The caveat "large enough" is deliberately and conveniently vague, but in the name of simplicity if not philosophy, most twentieth-century cosmological calculations assumed that the universe was absolutely uniform. The primordial nucleosynthesis calculations provide a classic example. Nevertheless, the universe is not uniform. On any scale. You have probably seen computer simulations of the large-scale structure of the universe, like the figure above, with long filaments resembling the interior of a lung or a Jackson Pollack painting.

The filaments are *galaxy superclusters,* the largest structures in the observable universe. Superclusters may contain hundreds

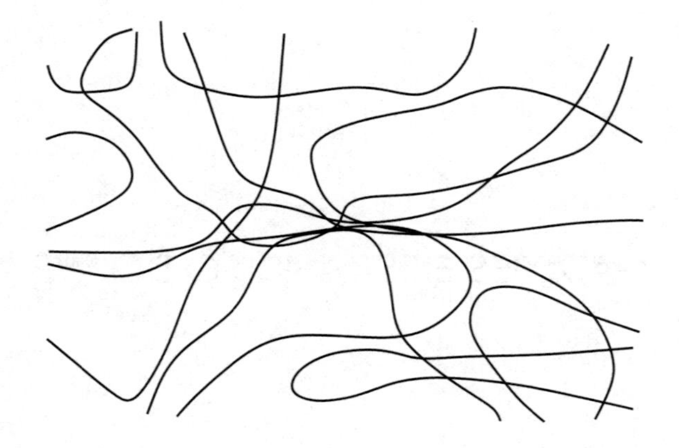

of thousands of galaxies and can be hundreds of millions of light-years long. The Milky Way is so small as to be invisible in this sketch.

Because the superclusters cannot be said to be randomly distributed in any strict mathematical sense, we are confronted by the inevitable question: How did the large-scale structure of the universe come into being? If the cosmological principle were exactly true, such a web would not exist, and neither would we. The fact of an irregular universe requires that the uniform big bang model be modified into a universe that, however uniformly it may have begun, quickly became otherwise. What's more, the standard model now must become one in which ordinary matter and radiation cede place to dark matter and energy.

*

The push to understand the large-scale structure of the universe has probably been the major focus of cosmology for the past four decades. Key to the entire endeavor has been the cosmic microwave background radiation. Although for three decades after its discovery the CMBR appeared completely uniform, cosmologists knew that, for galaxies as we know them to exist, they must have begun forming at the same time the observed background was created, 380,000 years BB, and their origins must have left faint imprints on the background.

When these traces were finally discovered by COBE in 1992, the popular press—and many prominent cosmologists—went celestial, announcing the discovery of the "fingerprints of God." To be sure, champagne was cracked by the COBE team, but cosmologists knew the situation would have been more interesting had the observations revealed nothing. Physics thrives when theories and observations clash—something, somewhere is wrong. In this case, the observations simply confirmed the theoretical predictions.

The theory of galaxy formation, which I'll use as shorthand for "large-scale structure formation," may be the finest example of the unity of cosmology: It demonstrates how precision observation, particle physics, and mathematical reasoning lead to a convincing picture of our universe.

*

At its simplest level, the process of galaxy formation is one of gravity versus expansion. Gravitational attraction attempts to clump matter into structures; the universe's expansion attempts to prevent it. Who, or what, wins?

To convincingly answer this question, let us first talk about sound. And to talk about sound, let us talk about Gaul. Like all Gaul, physics is divided into three parts: particles, springs, and waves. To a physicist, what is not a particle is a spring, and anything that is not either must be a wave. Newtonian physics is the physics of particles; modern field theories are the physics of springs and waves (the discussion of vacuum energy in Chapter 8 being a pertinent illustration). A true physicist quickly reduces any problem to one about springs, waves if required, or if speaking about galaxy formation, sound waves and light springs.

A sound wave, like any wave except light, is a disturbance traveling through some medium—say, air. A stereo speaker oscillates. The speaker's oscillations alternately compress the air in front of it and allow it to expand—or rarefy, as physicists say. Indeed, a small packet of air is compressed until the air pressure within the packet has increased enough to prevent further compression, and that pressure then causes the packet air to re-expand. When the packet pressure has dropped below the pressure of the surrounding air, the ambient air compresses the packet once again. Air is a spring.

Thus, the speaker has set up a series of oscillations, which propagate across the room. It is these oscillations that form the sound wave, as shown in the figure above, which travels at a velocity that depends on the ambient air density and pressure. In a typical room, the speed of sound is about 340 meters per second. The stiffer the material, the higher the speed of sound. The speed of sound in steel is not quite six kilometers per second, seventeen times higher than in air.

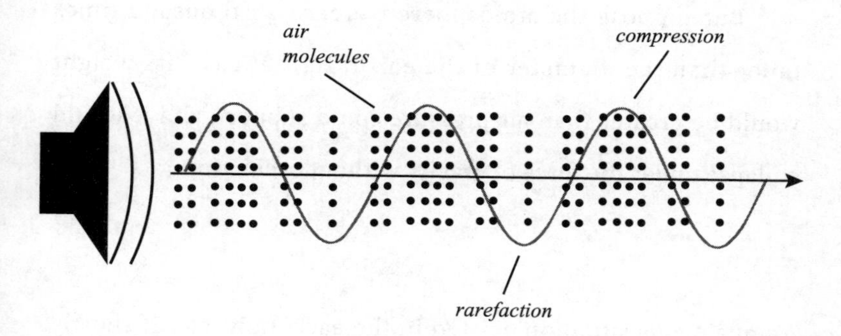

In a simple sound wave the air pressure or density oscillates from high to low in the pattern of a classic sine wave, as in the figure. The distance between any two adjacent pressure maxima or minima is the wavelength of the disturbance, which for audible frequencies is in the meter range.[①]

Let us move outdoors. The earth's atmosphere is a big room, one that would collapse under its own weight if not for the air pressure supporting it against gravity. In the real atmosphere, the air pressure is quite sufficient to prevent this from happening. Just as is the case indoors, if a tall air column in the atmosphere is compressed a little, the pressure builds up and forces the column to re-expand. It overshoots until the column pressure drops below the ambient air pressure, which forces the column to re-compress. Physicists say that the atmosphere is stable against gravitational collapse and merely undergoes "acoustic oscillations"—a fancy term for sound waves.

But suppose the atmosphere were, say, a thousand times taller than the diameter of the earth. In that case, its weight would be greater than air pressure could support, and it would collapse under the force of gravity without oscillating.

✳

An analogous situation existed in the early universe. If shortly after the big bang the primordial soup was spread uniformly

① See footnote on page 60.

throughout the universe, then the gravitational attraction of matter caused it to start clumping. Air pressure did not exist in the early universe, but light pressure did. In Chapter 5 we saw how, before the era of recombination, photons were unable to travel far before colliding with electrons. Photons striking matter exert a pressure on it, the same pressure that might allow sail-rigged spacecraft to cruise in the solar system under the pressure of sunlight. This pressure opposes the tendency of the matter to collapse under its own weight, and acoustic oscillations ensue, exactly as sound waves in air.

The first major difference between air in a room and light in the early universe is that the primordial soup was much stiffer than air. Steel, being stiffer than air, may have a sound velocity seventeen times higher, but the speed of sound in the early universe was nearly sixty percent the speed of light (for sticklers, $c/\sqrt{3}$). Consequently, the primordial construction material was so stiff that the *smallest* structure that could have collapsed was more massive than a supercluster of galaxies, which has a visible mass of about 10^{16} suns. In other words, no structures were formed in the very early universe.

Remember, though, the CMBR came into existence during recombination, when neutral atoms were formed, at which point the photons ceased striking the matter particles. That is equivalent to saying that the light pressure on the matter dropped to near zero, with the concomitant outcome that the primordial soup became much less stiff. As a result, much smaller structures could collapse—indeed, structures of about 10^5 solar masses, which is less than a millionth the mass of the Milky Way and about the mass of a globular star cluster.

Before photons and matter parted company at recombination, they essentially acted as one soup, so when matter began to clump, photons clumped along with it. These tiny variations in photon density manifest themselves in slight temperature variations of the CMBR. It is these variations that were the fingerprints of God discovered by COBE, measured with great accuracy by its successor satellite, WMAP (Wilkinson Microwave Anisotropy Probe), and measured with extraordinary accuracy by Planck. Although the fluctuations were only about a hundred-thousandth of a degree, they were precisely large enough to produce by gravitational collapse the structures we now observe. Today, the "bottom-up" collapse scenario provides the accepted picture of galaxy formation: the smallest structures formed first, and these gradually coalesced into larger structures. Superclusters of galaxies are forming even as you read this sentence.

Has something been left out of this picture?

THE UNIVERSAL PIPE ORGAN

THE ANALOGY of a few pages ago, comparing the universe to a room, omitted an essential difference: the universe is expanding. Because expansion pulls structures apart, it hinders gravitational collapse. The outcome of the competition depends on the exact expansion rate, which in turn depends on how much and on what ingredients are available.

Photons do not behave like matter, and dark energy does not behave like either, so it should not be too surprising that the expansion rate of the universe depends not only on the density of its contents, but also on the contents' nature. A universe of visible or dark matter (matter dominated, in the language of Chapter 5) expands at an ever-slowing rate. A radiation dominated universe, where photons or neutrinos are in charge, expands at a different ever-slowing rate. A universe filled with dark energy—ruled by the cosmological constant—increases its size at a constant expansion rate. A highly curved universe has a yet different behavior.

Since the expansion rate differs so much depending on the components, you might guess that changing their proportions alters the outcome of any galaxy-formation scenario. This is true. It's also fortunate, because it allows cosmologists to exclude most conceivable proposals. The question then becomes: What are the precise proportions of ingredients that permit galaxies to form within the current age of the universe?

*

In attempting to answer this question, let us turn again to sound, in particular to pipe organs. The dominant feature of a church organ is its ranks of hundreds of pipes of differing lengths. The length of an organ pipe determines the note it sounds. Specifically, the pipe length determines precisely what wavelengths or frequencies *resonate* within that pipe. Organ pipes come in many varieties, but some are essentially open at both top and bottom. As a sound wave travels through the pipe, compressing the air and allowing it to rarefy, the pressure at the open ends must remain the same as the pressure in the room. That is the condition for air to resonate within the pipe.

As illustrated on pages 115 and 116, the longest wave that can be put in such a cavity that meets this requirement is one with a wavelength twice the length of the pipe. That is the fundamental, or first, harmonic—the note we hear.

A wave whose wavelength exactly equals the pipe length also meets the resonance condition. Since its wavelength is half that of the fundamental, it has twice the frequency. This is known as the first overtone, or second harmonic. The third

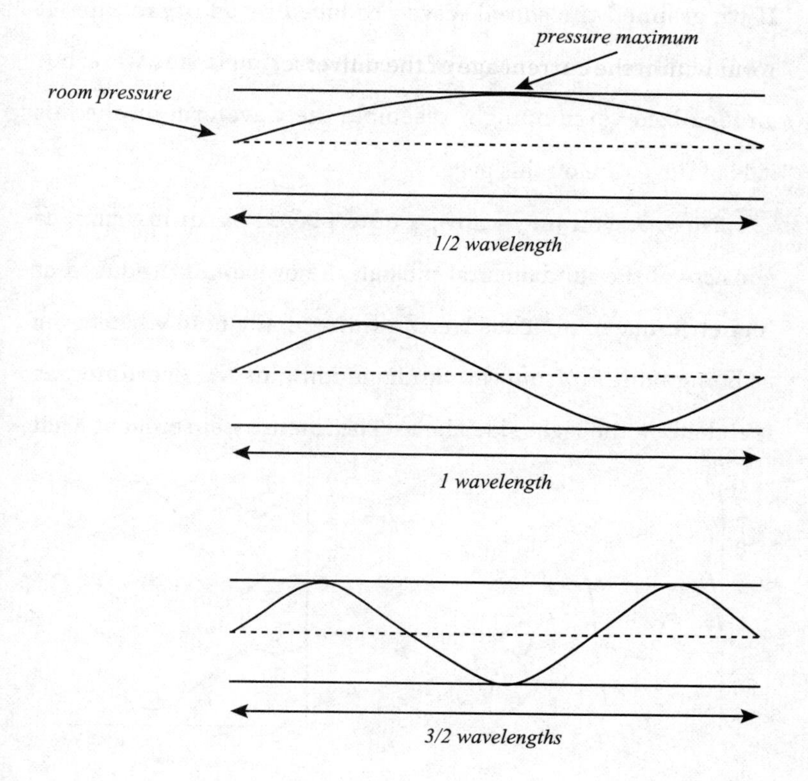

harmonic, which oscillates at a frequency three times that of the fundamental, also resonates, and so on. In all these cases, the distance from a pressure maximum or minimum to the nearest point of room pressure is one-quarter of a wavelength, or one-quarter of an oscillation.

Basically, the universe is a pipe organ.

✷

If we graphed the sound wave produced by an organ pipe, it would look much more complicated than a simple sine wave, but an idealized version might resemble the waveform on the left side of the figure on this page.

Now, as you may know, a note played by an instrument consists of the fundamental plus all the overtones produced at higher frequencies. Thus we can think of any note whatsoever as being built up from the fundamental plus the overtones, as sketched on the right side above. The intensity of sound at each

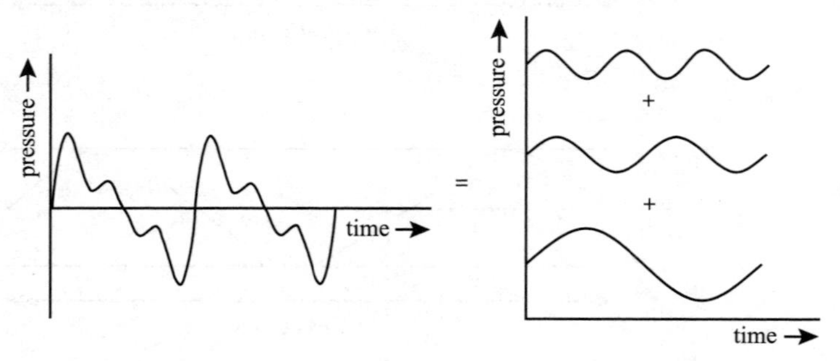

frequency determines the original note's shape. Mathematically, the technique used to break down a note into its overtones, or harmonics, is called *spectral analysis.* Having decomposed a wave into its harmonics, we can plot a graph like the one on this page, which shows the amount of sound energy at each frequency. This is a sound spectrum—the same as it was for light or heat. These figures depict a simple case containing only three harmonics.

The early universe was the grandest pipe organ conceivable. Bear in mind that the temperature fluctuations detected in the cosmic microwave background are proxies for fluctuations in the matter density. These fluctuations are not all of the same magnitude. The detailed map created by the Planck space telescope shows that some fluctuations display a higher density than others, resulting in a spectrum of density fluctuations

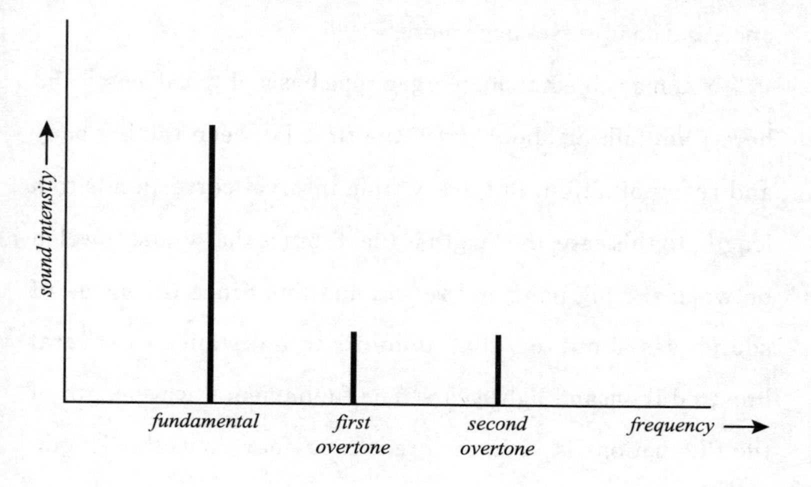

completely analogous to the sound spectrum of an organ pipe.

Indeed, the physical sizes of the density clumps are exactly determined by the resonant frequencies of the early universe. Imagine that shortly after the big bang, all matter is spread uniformly. It begins to clump, but light pressure forces the clumps to oscillate. The oscillations stop when the photons part company from matter at recombination. In the organ pipe, a maximum of pressure is one-quarter of an oscillation away from "ambient pressure," which in this case is the light pressure of the early universe. The *funda-mental* oscillation in the early universe is thus the one in which a clump of matter has had a chance to compress once from its starting condition until recombination, when oscillations cease. This first overtone compresses once and expands once. The second overtone compresses once, expands once, and compresses once more.

You may object that an organ pipe has a physical *length* and here I am talking about *time*—the time between the big bang and recombination. But every time interval corresponds to a length. In this case, the length is the distance that sound traveled between the big bang and recombination. Since the speed of sound was about .6*c*, that amounts to a distance of several hundred thousand light-years. The fundamental wavelength of the fluctuations is, as in the organ pipe, four times this length.

The wavelengths of the overtones are correspondingly smaller.

The universe has expanded by roughly a thousand times since these oscillations imprinted themselves on the background radiation. Because waves expand with the universe, the wavelengths of all the harmonics have stretched by the same amount, but they can be readily translated into separations as seen on today's sky. The fundamental should appear at an angular size of about one degree—twice the diameter of the moon. The overtones should appear at correspondingly smaller sizes.

Most extraordinary is that, in a series of ground-based and satellite observations spanning several decades, the predicted harmonics have been discovered. For instance, the Planck map showing the primordial density fluctuations can be decomposed into a sound spectrum. A graph of such *baryon acoustic oscillations*—sound waves to most people, fingerprints of God to enthusiasts—is shown at every cosmology seminar. As sketched on page 120, the first peak is the fundamental of the universal organ, the other peaks are the overtones.

Because clumping depends on the expansion rate of the universe, which in turn depends on its contents, this graph should reflect that. Indeed, the CMBR fluctuation spectrum has become one of the most sensitive tests of our cosmological

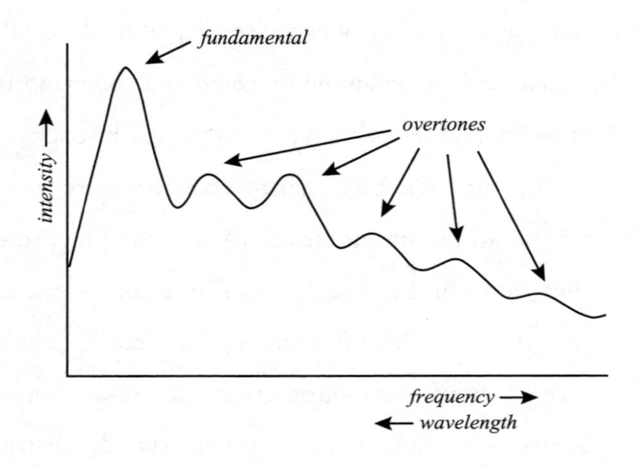

models. In a closed universe—one curved like a sphere—a distant object will appear larger than it would in flat space. This has the effect of shifting the peaks toward larger angular sizes, which on this graph is to the left. For the peaks to be exactly where they are observed, the universe must be, as far as anyone can tell, flat. This is the principal reason I stated in Chapter 3 that the geometry of the universe is nearly Euclidean—which is to say, flat.

If the universe is flat, then by definition the density of all its ingredients—ordinary matter, radiation, dark matter, dark energy—must sum to the critical density discussed in Chapter 4. That being the case, the great cosmological game is to juggle the proportions of the universe's constituents to give the best fit to the observed graph.

Take matter. If ordinary baryonic matter (neutrons and protons) were the only matter in the universe, it would have begun to clump only when the light pressure on it disappeared, at recombination. But by now you are convinced that most of the matter in the universe is dark, meaning more precisely that it does not interact with light in any way. Consequently, the light pressure of the early universe had no effect on it whatsoever, and it could not have engaged in any acoustic oscillations.

Dark matter does make its presence felt by gravity, and so it would naturally clump. Indeed, if dark matter consists of heavy WIMPs—say, one hundred times the mass of the proton—it must have begun clumping almost immediately after the big bang. Because the presence of dark matter would become appreciable at the time when the universe became matter dominated, as described in Chapter 5, which is earlier than recombination, it would have provided gravitational nucleation centers to nudge along the clumping of baryonic matter. More clumping translates into higher peaks in the primordial sound spectrum.

Suppose instead that dark matter consisted of neutrinos. Dark matter is dark matter, and in that sense neutrinos are no different than WIMPs, except that we know they exist. Neutrinos could therefore have provided the same sort of gravitational nucleation centers for baryons to give clumping a head start. The problem is that, compared to WIMPs, neutrinos are extremely light particles, streaming at nearly the speed of light in the early universe. This is much too fast to allow them to clump under their own gravity unless there were roughly a supercluster's worth of them—and in that case, the nucleation centers would be nearly the size of the universe and there would be no formation of small structures, like the globular clusters.

High-velocity particles are referred to as *hot dark matter,* in contrast to heavy, slow movers like WIMPS, known as *cold dark matter* particles. In general, the higher overtones in the acoustic spectrum, which represent clumping at smaller sizes, are washed away in hot dark matter universes. Because the higher overtones exist, cosmologists believe that dark matter in the universe is cold.

The cosmological constant, today's major ingredient in determining the expansion rate, turns out not to have a big effect on the CMBR spectrum. Although "outweighing" matter (visible and dark) in energy density *today,* it had the same energy density in the early universe—it is, after all, a constant. But the energy densities of matter and radiation rapidly increase into the past and would have overtaken the energy in the cosmological constant only a few billion years ago. Thus, the constant played little role at the formation of the CMBR, which was much earlier yet. Nevertheless, cosmologists believe it exists due to the acceleration of the universal expansion, and for other reasons I've so far left unmentioned.

One of these is *gravitational lensing of the cosmic microwave background*. Just as the MACHOs in Chapter 7 would distort the image of any light source behind them, the Planck map of the CMBR is distorted by any intervening matter—say, superclusters—lying between us and the edge of the observable universe, nearly fourteen billion light-years distant, where the CMBR was created. And just as the image produced by a magnifying glass depends on its position between the eye and the object, the distortion of the CMBR depends on the position of the lensing matter. In an expanding universe, that will depend on all the above ingredients, including the cosmological constant. Juggling the proportions to provide the best fit for the CMBR spectrum requires dark energy.

And so, at last, we arrive at today's standard cosmological model, usually abbreviated ΛCDM, for Lambda Cold Dark Matter. The best fit to the curve requires 68.5 percent dark energy, 26.7 percent dark matter, and 4.8 percent ordinary matter—but don't quote me.

*

As successful as the ΛCDM model is, it does leave open questions. First, once all the ingredients are in hand, it is reasonably straightforward to calculate the value of today's Hubble constant. Unfortunately, the value researchers find by considering baryon acoustic oscillations and gravitational lensing is about 67.4 in the standard units employed by astronomers, while the value determined by the supernova measurements is 73.9, a 10 percent discrepancy.[1] Astronomers pursue the Hubble constant with the zeal of crusaders, and so one can be sure that they will not rest easy until the matter is resolved.

Is a 10 percent discrepancy important? Observations of small deviations from Hubble's law did lead to the discovery of the universe's acceleration. In the present situation, however, a mistake somewhere along the line is more likely. Soon enough, measurements will reach a point—say, hypothetically, where the discrepancy is one percent—when further refinements in the value of H won't guide us to new physics, and it might be wise before reaching that point to ask what the aim of the pursuit is.

More importantly, I have not really talked here about

---

① Astronomers would write 67.4 kilometers per second per megaparsec.

structure formation, but only about the beginnings of structure formation. As the universe evolves, however, forming galaxies and stars, the physics becomes more complicated, because forces other than gravity come into play. For the record, for several hundred million years after the creation of the CMBR, the universe entered a "dark age." At the end of that period, the earliest galaxies made their appearance. Galaxies began grouping into clusters several hundred million years after that, and superclusters are still coming into existence today.

All these structures can appear within the age of the universe, assuming that the size of the fingerprints of God at the creation of the microwave background is what is observed: one part in a hundred thousand.

Moreover, the fingerprints of God spectrum has an interesting property, being what cosmologists term *scale invariant*. Loosely, scale invariant means that things look the same at any size. Zooming in on a fern leaf, you see that it appears very much the same in the small as it does in the large. Cartons of Land O'Lakes butter used to feature a Native American woman holding a Land O'Lakes carton, showing a Native American woman holding a Land O'Lakes carton, showing a Native American woman holding a Land O'Lakes carton. . . . If the sound intensity per octave in an organ-pipe spectrum never changed, we might say that the spectrum was scale invariant. If you prefer, call it the "Land O'Lakes spectrum."[①]

In the early universe, the clumping intensity compared to the clump volume remains constant. It is far from obvious that the spectrum produced by the baryon acoustic oscillations should be "Land O'Lakes," but it is.

What fixed the size and spectrum of the fingerprints of God?

---

① A more accurate definition would be to say that sound intensity per cubic wavelength per octave should be constant. In the case of the CMBR, "intensity" refers to the square of the amplitude of the density fluctuations.

THE FIRST BLINK: COSMIC INFLATION

UP TO THIS POINT, the story has concerned the universe after .0001 second BB, when primordial nucleosynthesis was soon to get underway. It is natural to wonder what happened at earlier times, but here things become more, let's say, speculative. Going back to about a microsecond BB, we expect that neutrons and protons would be boiled into their constituent quarks, and this behavior has been recently confirmed in earthbound particle colliders, but whether a plethora of altogether new particles makes its appearance at still earlier times is unknown. The Higgs boson would have existed in the first billionths of a second BB. The Higgs is the fabled particle that helps give mass to yet other particles, but I mention it only in passing because it does not play a central role in the cosmology plot. Clearly, thoughts of the dreaded singularity, when everything completely blows up at *t*=0, are beginning to intrude, but for the moment let us continue to avoid a direct confrontation and ponder the first instants after the big bang, as cosmologists do, despite all the uncertainties.

Just after 1980, a new theory of the first 10^{-32} second BB captured the imagination of the cosmological community—and soon thereafter, the public's imagination. For reasons that will become obvious, it went by the name *inflation,* a term coined by its principal protagonist, Alan Guth, who had been giving seminars on his idea, although similar proposals had already been published by Demosthenes Kazanas in the United States and Alexei Starobinsky in the Soviet Union.

For a number of reasons, not least the name, inflation took off. Almost at once it became incorporated into the standard cosmological model, textbooks presented it as a done deal, and four decades on, inflation continues to be a cornerstone of cosmological thinking. You should understand that inflation is not a theory in the standard sense of the term, like quantum mechanics, which has been verified by myriad experiments and observations. Rather, by now, inflation represents a collection of hundreds of models whose original purpose was to explain certain "defects" in the big bang theory as I have presented it. These are not observational anomalies but theoretical or philosophical conundrums that the standard big bang simply does not address. They are much closer to the photon-to-baryon puzzle in Chapter 6 or the cosmological constant problem in Chapter 8 than they are to the perihelion shift of Mercury. Whether inflation has truly solved these mysteries has become the subject of ever more heated debate, and whether it will emerge victorious or be relegated to the ash heap of history is for future cosmologists to determine.

*

Two problems inflation was invented to solve had been long emphasized by Robert Dicke; the first of them is known as the *flatness problem*. As maintained throughout this book, the real universe, as observations confirm, is very nearly flat. Why?

"Why not?" you might respond, but the matter is not so easily dismissed. If the present universe is nearly flat, the density is close to the critical value that divides the "closed" spherical universe from the "open" potato-chip universe in Chapter 4. How likely is this? To illustrate, suppose the density today is 99.5 percent of critical. It is then easy to show that at one second after the big bang, the start of element formation, the density would have to have been within one part in 10^{17} of the critical value, and at 10^{-36} second BB, a time I have not chosen at random, it would have to have been flat to one part in 10^{52} or so. In other words, the universe would have to have been fine-tuned to flatness with unimaginable precision.

Even those inclined to accept the occasional coincidence find it totally improbable that the big bang could have been so flat. As with the photon-to-baryon ratio and cosmological constant conundrums, this is very much a *why* question. As before, cosmologists find it vastly preferable to transform it into a *how* question—they'd prefer to avoid any fine-tuning and find a mechanism to drive the universe to flatness, regardless of how it began.

But what does "probable" or "improbable" mean when only a single universe is at our disposal? Here we run full force into the difficulty posed by the uniqueness of the cosmos. We grapple with it in the next chapter.

The second of Dicke's conundrums that inflation claimed to solve is known as the *horizon problem*. The temperature of the CMBR is observed to be remarkably uniform in all directions. Even the "fingerprints of God" of the previous chapters change the uniformity by only the thickness of a marble compared to the height of the Burj Khalifa, the world's tallest building. How did this remarkable uniformity come about? Another coincidence?

Perhaps, but to make the situation more vivid, let's say that in the observable universe there are 10^{87} photons, a large number. Because they are within the observable universe, they are within the distance light has traveled since the big bang—the *cosmological horizon* discussed in Chapter 4. Because no signal can travel faster than light, the cosmological horizon provides the ultimate communication barrier: no two objects can influence each other in any way if they lie beyond each other's horizon. As shown above, A's horizon lies at the distance light has traveled since the big bang, *(speed of light)* × *(age of universe)=ct*. *A* and *B*, lying within this distance, can have influenced each other. *A* and *C* cannot influence each other until the horizon has grown to the distance between them. *A* and *B* are said to be in *causal contact,* while *A* and *C* are not.

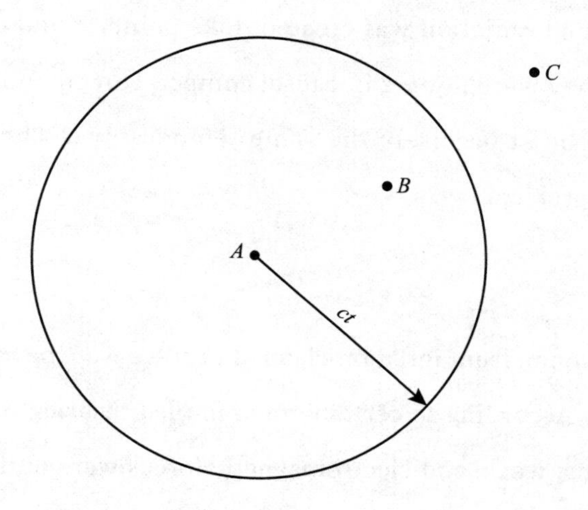

By definition, everything within today's observable universe lies within the cosmological horizon. Also by definition, the horizon grows at the speed of light, and therefore, going back toward the big bang, it shrinks at the speed of light. On the other hand, the universe's expansion rate—the rate at which galaxies are receding from one another—is less than the speed of light. Therefore, going back into the past, the universe shrinks more slowly than the horizon. Consequently, as we approach the big bang, the universe within the horizon occupies an ever-smaller fraction of what became today's observable universe. At the time the CMBR was created, only about a hundred-thousandth of today's universe lay within the horizon—say, 10^{82} photons.

This means that two widely separated patches of CMBR photons could not have spoken to each other at the time the background radiation was created. Like points *A* and *C* in the figure, they were not yet in causal contact. How then did they come to be at precisely the same temperature? That is the horizon problem.

A third conundrum inflation claimed to solve was the *monopole problem*. According to certain *grand unified theories,* or GUTs, the strong, weak, and electromagnetic forces were unified into one grand unified "field" at the enormously high temperature of 10^{29} degrees that occurred at about 10^{-37} second BB. As the universe expanded and the unified field split into the individual fields, socalled magnetic monopoles were produced. A magnetic monopole would be an isolated north or south pole of a magnet, the magnetic analog of positive or negative electric charges. But although isolated positive and negative charges are found everywhere as protons and electrons, no one has ever observed an isolated north or south magnetic pole. All magnets have both a north and a south pole and cutting the magnet in half merely produces two smaller magnets, each with its own north and south pole.

Nevertheless, some GUTs predict that magnetic monopoles should have been produced in copious numbers in the early universe, and they would be so heavy (sixteen orders of magnitude heavier than the proton) that they would completely dominate the density of the universe. That is the monopole problem.

*

Inflation's solution to all three of these problems was elegant and straightforward enough that the average physicist could understand it. It postulated that as the GUT era ended—say, between 10^{-36} and 10^{-32} second BB—the universe underwent an enormous spurt of exponential expansion, increasing its size by twenty-seven or twenty-eight orders of magnitude in that incredibly short amount of time. This is equivalent to blowing up a popcorn kernel to the size of the observable universe.

If you were an ant walking on the surface of a popcorn kernel that suddenly inflated by twenty-seven orders of magnitude, its surface would appear exceptionally flat. This is inflation's solution to the flatness problem.

The monopole problem goes away in the same stroke. The vast numbers of monopoles in the universe were simply diluted by the enormous expansion so that their density became about one monopole per observable universe, and we haven't found it.

The horizon problem is more involved. It asks how it is that widely separated parts of the sky could have interacted and smoothed each other out to produce a uniform microwave background. Because in the standard model the horizon shrinks toward the past faster than the size of the universe, the horizon at 10^{-36} second was smaller than the size of the universe by about twenty-seven orders of magnitude. Thus, virtually none of the particles in the universe could interact. On the other hand, by definition, the particles within that tiny horizon could have communicated. If that patch inflated by twenty-seven orders of magnitude, it would now be the size of the observable universe.

This is what inflation claims to have done: it posits that the present universe grew out of a popcorn kernel-sized patch of sky in which the photons had already interacted and smoothed out any irregularities; inflating it would produce a uniform background radiation. Note, however, that inflation does not explain *how* the smoothing took place; it only provides the necessary condition that the smoothing *could* have occurred.

A principal reason that inflation became so popular had nothing to do with these three conundrums, but with the fingerprints of God. The fluctuations in the microwave background represent temperature changes of one part in a hundred thousand compared to 2.7 degrees. They also display the Land O'Lakes spectrum, scale invariance. Both these features are observational results. How did they arise?

Early inflationary models claimed to account for them. Recall from Chapter 8 that physicists believe the vacuum of space is filled with small energy fluctuations, the so-called vacuum energy fluctuations. Inflation posits that these quantum fluctuations existed immediately after the big bang, produced in the era of quantum gravity, which will appear in Chapter 14.

Inflation takes these fluctuations and, well, inflates them, until they become the fluctuations in the CMBR. What's more, it does so in such a way that the spectrum of these oscillations is Land O'Lakes.

*

So, if inflation occurred, it could apparently explain certain puzzling features of our cosmos. But how did inflation itself come about? This is where the hundreds of different inflationary models differ. Most posit a new field, not unlike dark energy. Remember, the expansion rate of the universe depends on its contents. If the universe is dominated by dark energy—a cosmological constant—then Friedmann's equations say the size increases *exponentially* with time. In fact, because today's universe is dominated by a cosmological constant, it is now expanding exponentially, approximately.

In the inflationary scenario, much the same took place between 10^{-36} and 10^{-32} second BB. At that time, the universe was dominated by a new form of energy, which was not necessarily the dark energy of today, but resembled a cosmological constant for a time, as sketched on page 137. This nearly constant energy produced inflation's exponential expansion and, at the end of the inflationary period, decayed away until it disappeared. This figure is known as a potential energy diagram. As you may know, any system, like a ball on a hill, tends to seek the lowest energy, which is why balls roll downhill. Physicists often visualize the universe itself as a ball sitting on top of the energy curve provided by the inflationary field. As the ball slowly rolls down the almost flat hill, inflation takes place. At the end of inflation, the ball rapidly plunges into the well, losing all its energy.

Physicists, however, also subscribe to the famous law of conservation of energy and are reluctant to believe that the dominant form of energy in the universe vanished without a trace. Rather, the basic picture is this: during inflation the universe expanded enough to solve the cosmological conundrums. The enormous expansion also utterly emptied the universe of all its contents—monopoles, photons, neutrinos, and anything else. When inflation ended, the field driving inflation

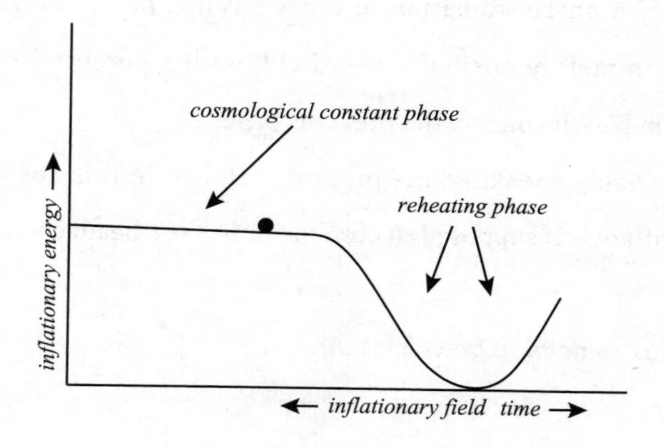

decayed away, transforming its energy into the particles that make up the present universe. Inflation plus the subsequent "reheating," as it is called, all happened in far less than the blink of an eye.

Why does the inflationary energy decay away? The original proposal was based on the wellknown phenomenon of *phase transitions.* Water can be cooled far below freezing when done slowly and carefully, but if a dust particle finds its way into the water it becomes a nucleation center for ice, and the water rapidly freezes everywhere. In the context of grand unified theories, it was plausible to think that something like this happened with the vacuum energy of space in the early universe as the unified forces fragmented into their distinct identities. The vacuum energy began at a large value, then became "supercooled," during which time inflation occurred and finally suffered a phase transition to today's value. Later versions of inflation merely posited a new field, with a potential energy diagram like the one on the previous page.

Roughly speaking, the picture sketched in this chapter is how inflation is supposed to cure the universe's headaches.

Has something been left out?

TO INFLATE OR NOT TO INFLATE

IN THE PREVIOUS DISCUSSION, I oversimplified—and even lied. While the inflationary picture provides an elegant solution to the famous cosmological puzzles, it has come under increasing scrutiny, as is proper in science, and today its future looks much less assured than it did in the years immediately following its advent.

Consider the monopole problem. Despite efforts spanning decades, no experimental evidence for GUTs has ever been found, and it may be that the theories predicting copious numbers of monopoles are simply incorrect, in which case the monopole problem vanishes.

Consider the fingerprints of God. Most accounts, popular and technical, focus on the spectrum of the fluctuations, and how that spectrum is in accord with the simplest inflationary predictions. Still, the size of the fluctuations—that hundred-thousandth of a degree—must also be explained. It has long been recognized that reproducing this number in simple models requires adjusting the shape of the potential in the diagram on page 137 to extraordinary precision—as in, change it by a part in 10^{14} or so and you get the wrong answer. This is another example of fine-tuning and it forces us to ask whether in choosing the potential to have the necessary shape, we have merely swapped one fine-tuning problem for another.

Furthermore, while the Land O'Lakes spectrum may be in accord with inflation's predictions, inflation is not the only process that can produce one (as we will see in the next chapter). If true, how does one decide among models? In fact, inflation actually does not predict a Land O'Lakes spectrum, only a nearly Land O'Lakes spectrum. At least a few cosmologists argue that Planck satellite data are already in conflict with inflation's true predictions and the theory should be discarded on observational grounds in favor of models to be discussed in Chapter 13. Needless to say, proponents of inflation disagree.

*

The inflationary scenario presents cosmologists with yet other ambiguities and difficulties. For example, as has been known for over two centuries, light reflected off a windowpane is *polarized.* What does this mean? Light, an electromagnetic wave, is composed of an electric and a magnetic field oscillating at right angles to one another as the wave travels. The direction in which the electric field points is termed the direction, or axis, of polarization. Light from an incandescent bulb is *unpolarized,* meaning that the bulb emits light whose electric field is randomly pointed in all directions. Unpolarized light can be thought of as consisting of two independent light beams whose electric fields are oriented perpendicularly to one another. When such a beam strikes a window, one direction is preferentially reflected by the glass and so becomes polarized—its electric field is oscillating in one direction only.

You know this to be true. Polaroid sunglasses work because their molecules are aligned in such a way that they transmit only one direction of polarization, thus cutting the intensity of unpolarized light in half. Because light reflected off a car windshield is already polarized, if you rotate your sunglasses until their polarization axis is at right angles to the light's electric field, you see almost nothing.

The cosmic background radiation is a large car windshield. At the time the CMBR was being created, photons were striking electrons, and this set them oscillating in the direction of the light's electric field. Because shaking electrons reemit light preferentially in one direction, the light is polarized. If the primordial soup were completely uniform, photons would strike electrons equally from all directions and the overall polarization would be zero. But the tiny fingerprints of God mean that the CMBR windshield is not exactly uniform—and this results in a small, net polarization.

The polarization of the microwave background has been precisely measured by many extraordinarily sensitive telescopes—too many to list, all with acronyms like DASI and ACT, based at the South Pole or in Chile's Atacama Desert—and all verify this picture.

Now, inflation *also* predicts the existence of primordial gravitational waves, produced by fluctuating quantum fields in the very early universe. Back in Chapter 3, we encountered the gravitational waves that travel across spacetime, tidally stretching and shrinking any detector set up to measure them. They did the same to the primordial soup as the CMBR was created, producing irregularities that also polarize light. The stretching and compressing of the background by gravitational waves produce a fingerprint that differs, however, from that produced by clumping due to acoustic fluctuations (the clumping discussed in Chapter 10). In principle, with a sensitive enough telescope, the two different patterns can be distinguished.

The polarization of the CMBR due to primordial gravitational waves is predicted to be far less than that due to acoustic fluctuations, but some cosmologists maintain that if such polarization is discovered, it will provide a "smoking gun" for inflation. Despite a very public announcement at Harvard in 2014 by the BICEP2 team of just that discovery, the results were eventually retracted and to date primordial gravitational waves have gone undetected. As already mentioned, some cosmologists say that the Planck satellite data already rules inflation out.

The main objections to inflation, however, spring from its fundamental assumptions. Although I have mentioned quantum fluctuations a few times already, and what inflation is supposed to do to them, it is important to understand that a quantum theory of the beginning of the universe does not yet exist. Inflation, then, cannot be a genuine quantum theory of the universe; rather, inflationary models use ordinary, classical physics to "mock up" presumed quantum behavior. Indeed, a major objection to inflation is that its fields have been introduced solely for the purpose of producing inflation, and have no observational or theoretical justification.

A related difficulty is the fact that inflation is meant to stretch the presumed primordial quantum oscillations until they become the fluctuations observed in the CMBR. No mechanism has yet been provided for the transition from the quantum to the classical theory. Indeed, if inflation proceeded a bit longer than necessary to solve the cosmological puzzles, then one can show that at the onset of inflation the wavelengths of the oscillations were less than 10^{-33} centimeters. This is a small number. In fact, this length, termed the *Planck length,* is the length at which physicists believe classical physics must break down altogether and below which a quantum theory of gravity must take over. Because no such theory yet exists, one must regard anything that relies on statements of what might have happened during the epoch of quantum gravity with skepticism.

For the moment, however, assume that the inflationary models sensibly reproduce quantum behavior. Quantum fields fluctuate randomly throughout the universe. Small fluctuations far outnumber large ones; nevertheless, large ones occasionally occur. During inflation, a large fluctuation in one place in the universe may move the field higher up on the curve drawn on page 137, leading to more inflation in that region before it ends. As that "bubble" inflates, more fluctuations will occur, producing daughter bubbles of longer inflation, *ad infinitum.*

Inflation is eternal, literally. One therefore ends up with a very irregular situation, with inflation occurring in different amounts in different daughter universes. In some places inflation may have solved the cosmological conundrums, but in other places it has not. This *multiverse* seems to be an inevitable consequence of the inflationary paradigm, and we will consider it more fully in Chapter 15.

For the moment, the important point is that the multiverse, while extremely popular with the public, presents extreme conceptual difficulties. Suppose we tried to estimate the probability that a given universe would solve the cosmological problems. If we are dealing with an infinite number of universes, this is, to say the least, tricky. When we throw darts randomly at a dartboard that is 25 percent yellow and 75 percent black, our intuition tells us that we should hit a black sector three times as often as a yellow sector. Even faced with an infinitely large dartboard, we still feel that we should hit black three times as often as yellow, and we can indeed define probabilities in a way such that this remains true.

On the other hand, if the dartboard contains an infinite number of unique colors, then the probability of hitting any one of them is essentially zero. Suppose there are an infinite number of greens, representing all the conditions that inflation can successfully handle, but also an infinite number of reds, yellows, chartreuses, and so on. Is the probability of hitting some shade of green now greater than zero? As with the black and yellow dartboard, we would need to be able to say something like *on a finite-sized dart-board, hitting green is three times more likely than hitting purple,* and then assume this remains true even on an infinite dartboard.

Inflation presents us with this dilemma. If you ask the probability of producing a universe that solves the cosmological conundrums, you need to decide which conditions—colors—are more likely than others, and there is simply no agreed-upon way of doing that. Cosmologists Gary Gibbons and Neil Turok have concluded that most universes do not inflate enough to solve the conundrums. Mathematician Roger Penrose has gone further. The equations of inflation are exactly like those of Newton in that, if you know the present state of affairs, you can predict the future, or reconstruct the past. If you imagine a very irregular and curved universe today—one far more irregular and curved than observations permit—and run the equations back to the pre-inflationary period, you will have produced a set of conditions that, by your own construction, inflation cannot smooth out or make flat. What's more, Penrose argues that such irregular initial conditions are unimaginably more probable than smooth conditions, which leads him to conclude that inflation cannot be invoked to produce a universe resembling our own.

*

A different sort of resolution to the cosmic conundrums has frequently been proposed. One might argue that only nearly flat universes permit life to evolve. If they are too closed, they almost immediately re-collapse into a big crunch, eons before galaxies have the opportunity to form. If they are too open, galaxies are also unable to form. Therefore, out of all the possibilities resulting from the multiverse, we must observe our cosmos to be as it is, because undeniably we are here. This is another example of anthropic reasoning (about which more will be said in Chapter 15). Physicists tend to be skeptical of such arguments because there is no way to conclusively test them, but they illuminate the severe difficulties introduced by the inflationary picture, given that we have only a single universe at our disposal.

An even simpler illustration of the dilemma arises from the fact that today's universe is dominated by dark energy. If this energy really is a cosmological constant that remains constant, then as the universe continues to expand, its matter and radiation content will be diluted until only the constant remains. Even the energy provided by the curvature of space will eventually vanish—and so such a universe becomes flat. Will cosmologists of that distant epoch say there is no flatness problem, because the cosmological constant provides a mechanism to make it flat? Will they say that, because the universe's flatness depends on the size of the cosmological constant, the flatness problem is really the cosmological constant problem?

Or will all the stars in the universe by then have died out, leaving no cosmologists to ask the question?

Are there alternatives to inflation?

CRUNCHES AND BOUNCES

HERE, CLOSING in on *t*=0, you are asking: "What happened before the big bang?" Or perhaps: "Was there a big crunch before the big bang?" Indeed, maybe it was you who came up to the podium in the Introduction to pose this after-lecture question, one even more popular than "Are we at the center of the universe?" or "What is the universe expanding into?"

The question of what happened before the big bang is a natural one and cosmologists have been pondering it since the discovery of the expanding universe. Many have been the proposals but there is still no definitive answer. Cosmologies in which periods of expansion alternate with periods of contraction are known as cyclic universe models, or "bouncing" cosmologies, and in the past decade they have begun to be taken seriously again as alternatives to cosmic inflation.

The concept of a cyclic universe is extremely attractive because it allows us to avoid thinking about a universe suddenly popping out of nothing at a definite moment in the past. Mathematically, this means we don't need to specify the conditions at the beginning of the universe because there is no beginning. But imagining a universe that oscillates forever between expansion and contraction is not easy, either.

The difficulty faced by cyclic universes has always been the *big-bang singularity.* We can no longer put it off. In the Friedmann cosmology, at the instant of the big bang, the temperature, pressure, density, and expansion rate of the universe all become infinite. This is an utter breakdown of the system as we understand it—far more serious than a plague or economic depression, either of which eventually end. At the big bang all the equations of relativity go up in flames and we simply do not know what happened before, and perhaps never will. Friedmann himself recognized that Einstein's equations permitted an oscillating universe but paid no attention to the singularity. When in the early 1930s physicist Richard Tolman created a more detailed cyclic universe model, he recognized the severe difficulty posed by the singularity, but assumed a miracle occurred, allowing the universe to reexpand after the big crunch.

*

For decades, cosmologists believed that more irregular universes than Friedmann's might avoid the singularity. Remember, the matter in Friedmann's model is distributed uniformly and if the universe is closed, space is spherical. In a contracting universe, as in a contracting ball, all the matter approaches the looming singularity equally from all directions, eventually producing an infinite density as everything within sight is crunched together at the same time into a single point. One can, however, imagine a universe that is not so symmetrical—perhaps one shaped like a cigar. In such a universe, matter would collapse faster in one direction than another, and conceivably the singularity would be avoided.

Unfortunately, this turns out not to be the case, and all attempts made along these lines failed. The singularity remained. Essentially the failure comes about because gravity is an attractive force, which focuses matter to a point regardless of irregularities. Powerful singularity theorems by Amal Kumar Raychaudhuri, Roger Penrose, and Stephen Hawking, dating from the 1950s through 1970, prove that under fairly general conditions a big-bang singularity is unavoidable.

But all theorems make assumptions, and the big-bang

singularity can be evaded by introducing a large enough repulsive force. The cosmological constant—dark energy—accelerates galaxies away from each other, providing exactly the sort of repulsive force necessary to dodge the singularity. The main questions are these: How big should it be to produce a big bounce without interfering with astronomical observations? And should it really be constant?

For instance, suppose the current expansion of our universe was preceded by a collapse. In the collapsing phase, the CMBR would be heating up and one might postulate a cosmological constant large enough to bounce the universe before it reached a temperature of one billion degrees, which would take place three minutes before the big crunch. However, after the bounce—our bang—no primordial nucleosynthesis would take place and, unless the light isotopes already existed in their current abundances, they would never be created. What's more, such a large cosmological constant would cause the universe to expand so rapidly that galaxies wouldn't form. Adding a simple cosmological constant to cure the Friedmann model of its singularity is not a viable option.

The trick, then, is to introduce something that resembles a cosmological constant at the beginning of the universe—perhaps like the potential energy diagrammed on page 137—which then disappears before it causes havoc. Numerous proposals have been made, all differing in their features and motivations, and we will not go into the gory details. An attractive option is to bounce the universe before it contracts to the Planck size of 10^{-33} centimeter, mentioned in Chapter 12, which would be reached at the Planck time, 10^{-43} second before the big crunch.

The Planck length and time mark the end of physics as we know it. At smaller lengths and shorter times, our usual conceptions of space and time probably break down altogether and a theory of quantum gravity is presumed necessary to describe the singularity or get through it. Quantum mechanics can indeed produce repulsive forces that might do the job, but as already mentioned, a theory of quantum gravity does not exist. If instead, a bounce occurs well before the Planck scales are reached, then there is no need to invoke quantum mechanics. In that case, we can rely solely on conventional physics, which does exist.

*

In the past decade, some bouncing cosmologies have exploited these precepts. Like inflation, they invoke a new field resembling a cosmological constant that causes a bounce, but in which the blessed event takes place at a time of about 10^{-35} second BB.

That is a long time (in physicists' minds) before the Planck era is reached; it is even before the GUT era is attained, in which case classical physics should be entirely adequate.

You should be wondering whether such models can solve the cosmological conundrums that inflation was designed to explain away. As it happens, some of them can, and in much the same way.

To understand how, first realize that the instant explanation I gave in Chapter 11 for inflation's solution to the flatness problem—that the universe merely inflated twenty-seven orders of magnitude in a blink to make it appear flat—was a lie (although one commonly perpetrated by cosmologists). If we stand on the beach, looking out over the ocean, the earth appears flat to us precisely because our horizon is only a few kilometers away, which is far smaller than the size of the earth. But if we were standing atop a mountain whose height was comparable to the radius of the earth, we would clearly see the earth's curvature.

So flatness is relative; you must always compare the distance to the horizon with the size of the earth. If the horizon is much smaller than the radius of the earth, the earth appears flat. Similarly, in Chapter 11 we saw that in a collapsing cosmos the horizon always shrinks faster than the universe does, so the universe looks ever flatter toward the big bang.

The same applies in bouncing cosmologies. As we approach the big crunch in a collapsing universe, the universe appears flatter and flatter because we see only smaller and smaller distances. It is this little, flat piece of spacetime territory that becomes our present universe after the bounce.

The horizon problem goes away in the same stroke. If you imagine the universe in the dim past, just as it began to re-collapse in the previous cycle, all parts of that universe are already able to communicate because they lie within the horizon. As the universe shrinks toward the crunch, the horizon shrinks faster and it is the small patch within the horizon that becomes the present universe after the bounce, as it did in inflation. Since all particles in the patch already communicated before the bounce, there is no longer a horizon problem.

One striking feature of modern bouncing cosmologies is that these problems can be solved by a very slow contraction, such that the collapsing phase does not necessarily mirror the expanding phase in reverse. In some models, the universe does not even have to contract much to do the job. Moreover, as hinted in the previous chapter, an exponential expansion is not the only mechanism that can produce the Land O'Lakes spectrum in the microwave background. Mathematically, the slow contraction of some models does exactly the same thing.

Also, do not forget that the primordial gravitational waves predicted by inflation but not yet discovered are assumed to be the result of fluctuations created during the epoch of quantum gravity. Because in bouncing cosmologies that epoch is never attained, essentially no primordial gravitational waves are produced. The multiverse, the unruly offspring of those quantum fluctuations, is not produced either.

Bouncing cosmologies are currently an active field of research, but history teaches us that active areas of research may find themselves abandoned in the blink of an eye. So, while it is early to decide whether a big bounce will cure the conceptual headaches induced by inflation, in this blink of an eye they do appear to be an attractive and viable alternative.

How does one know whether such theories are true?

WHY QUANTUM GRAVITY?

WE HAVE ARRIVED at 10^{-43} second after the big bang. It is time—if time means anything—to create a theory of quantum gravity. Should bouncing universes turn out to be unviable in avoiding the singularity, cosmologists will have no other option. The main drive to create a theory of quantum gravity, however, is not so much the singularity itself as physicists' centuries-old conviction that the forces of nature should be unified into one towering edifice, the legendary *unified field theory*.

No observation ever made contradicts general relativity, and it is therefore considered to be as correct as a scientific theory gets. Yet it is a classical theory, taking no account of quantum phenomena. Modern quantum field theories have been tested to about the same precision as general relativity—arguments persist over the winner—but they take no account of gravity.

Theoretical physicists are convinced to the marrow of their bones that these two very different species should be joined into a consistent quantum theory of gravity. Nearly a century of effort, however, has gone into arranging a marriage without success. On the coarsest level, the difficulty has been that general relativity is a theory of the very large, while quantum theory is a theory of the very small. That clarification is unlikely to satisfy, but as physicist John Wheeler once remarked, the most difficult question about quantum gravity is: What is the question?

Let us ask a few basic questions, then; expect no answers.

First, what are quantum phenomenon? And at what point should quantum mechanics and relativity be wedded? The word *quantum* has long been part of our popular vocabulary, but despite the efforts of automobile branders and quantum healers, its exact meaning may remain fuzzy. In classical physics, most properties of a system—its energy for example—are permitted in any amount. The basic precept of quantum mechanics is that, no, these quantities come in discrete, or *quantized,* units, just as cash comes only in integral multiples of pennies. When Max Planck created quantum mechanics in 1900 by explaining the very black body spectrum of Chapter 5, his fundamental postulate was that the light emitted by the black body was quantized such that its energy equaled only integer amounts of the light frequency multiplied by a new constant of nature, which he labeled *h*. This number, now universally called *Planck's constant,* fixes the size of all quantum phenomenon.

In 1905, Einstein showed that not only was light quantized in Planck's sense, but that light should actually be associated with packets of energy, or *quanta,* which behave as particles. When Planck talked about the black body emitting light, he really meant light quanta—photons. A photon's energy is given by the light frequency multiplied by *h*. Swarms of photons acting in concert constitute a light wave, and when we study waves we no longer pay attention to the properties of individual quanta. A light wave is described by Maxwell's classical theory of electromagnetism.

One way of saying that a theory is quantum is that *h* is in there somewhere. If a theory doesn't contain *h* it is a classical theory. You won't find *h* in general relativity no matter how hard you look. On the other hand, being a classical theory of gravity, every one of its equations features Newton's gravitational constant *G,* which determines the strength of the gravitational force.[①]

The second important feature of quantum mechanics

---

① See footnote, page 14.

involves a famous phrase: *wave-particle duality.* Just as light can behave as particles, particles can behave as waves. Every particle has wave properties associated with it. In particular, it has a wavelength, which depends on the particle's mass and its velocity—and on *h.* Think of this wavelength as the quantum size of the particle, its size when it is behaving like a wave. For subatomic particles, like electrons, the wavelengths tend to be very small, roughly the size of an atom, and are unnoticeable in everyday life. In systems of atomic size, however, as inside modern electronics, the wave nature of matter becomes extremely important.

*

With these concepts we can understand the scales at which general relativity and quantum mechanics should be joined—precisely, the Planck mass and the Planck time of the previous chapters. You may know that any unit of measurement, be it metric, English, or potrzebie, is based on three fundamental quantities: mass, length, and time. The question is, what is the most sensible way to choose these three basic quantities?

In the nineteenth century, physicist George J. Stoney argued that it was better to base units of measurement on naturally occurring quantities, such as the electron's charge, the speed of light *c*, and the gravitational constant *G*, rather than on the length of a stick in Paris. Later, Max Planck had the same thought and proposed that the fundamental constants *G*, *h*, and *c* be made the basis for a system of units, today called natural, or Planckian, units. With a little patience you can combine *G*, *h*, and *c* into a length, which is about 10^{-33} centimeter, a time, which is about 10^{-43} second, and a mass, which is about 10^{-5} gram.[①]

Clearly, the Planck length and time are unimaginably smaller than anything you (or most physicists) would ever contemplate, while the Planck mass is unimaginably large compared to the mass of subatomic particles—large enough to be measured on a modern balance. If you multiply the Planck mass by c^2, you get the Planck energy, which is about 10^{15} times higher than the energies produced in the Large Hadron Collider, the most energetic particle accelerator on earth.

What do these bizarre numbers signify? The fundamental constants are the most important numbers in the universe because they determine the domain of all natural phenomena. *G* sets the strength of the gravitational force, while *h* determines when quantum effects are significant. When *c* appears in a situation, it shows that relativity is important—something is

① Specifically, the Planck mass is $m_p=\sqrt{hc/G}$; the Planck length is $\ell_p=\sqrt{hG/c^3}$; the Planck time is $t_p=\sqrt{hG/c^5}$.

moving near the speed of light.

You probably know that a black hole is an object whose gravitational field is so strong that light cannot escape; its size is given by its mass and *G* and *c*, nothing else. The size of a black hole can be thought of as the scale on which gravitational effects become extremely important. If you ask for the mass of a particle whose quantum size—its wavelength—is the same as its gravitational size, you get the Planck mass. The size of this quantum black hole is the Planck length, and the time for light to cross it is the Planck time.

So, the Planck scales represent the lengths, times, and energies at which quantum effects and gravitational effects are equally important. At these scales we cannot ignore either gravity or quantum mechanics and must create a quantum theory of gravity to describe the universe.

*

Why has such a theory proved so difficult to create? Fundamentally it is because the basic assumptions of general relativity and quantum mechanics are so different. Quantum mechanics ignores gravity and general relativity ignores quantum mechanics. Put another way, quantum theories assume spacetime is always flat, as in special relativity. General relativity assumes that spacetime can be curved, depending on its matter content.

This is a serious problem, which results in extraordinary technical difficulties. As originally created, quantum mechanics was, like Newtonian physics, a theory of particles. And like Newtonian mechanics, it took no account of special relativity. Wedding quantum mechanics and special relativity into *relativistic quantum mechanics* was accomplished by Paul Dirac in the late 1920s.

Relativistic quantum mechanics, however, continued to concern itself with particles—in particular, with electrons, which are regarded as point particles. Points, by definition, have zero extent. This produces the serious difficulty that when two point electrons touch each other, the electrical force between them becomes infinite.[①] Similarly, the energy of a point electron's field becomes infinite as one approaches the electron, and therefore so does its mass, which by $E=mc^2$ must include the energy of the field.

The efforts to resolve these dilemmas led to quantum field theories. In particular, *quantum electrodynamics* became the

---

① The electrical force between two point particles looks just like the law of gravity (see footnote, page 14), except that the masses are replaced by the electric charges and G is replaced by another constant. As the distance r between the two particles goes to zero, the force becomes infinite.

theory that explained how electrons interacted with photons. The naive hope was that, by smearing things out into fields, we need never get too close to point electrons, and such infinities—such singularities—would disappear.

A little less vaguely, in quantum field theory all interactions are described by exchanges of particles—the electromagnetic force is really due to an exchange of photons. Such exchange particles are termed *virtual.* We can regard them as manifestations of the vacuum fluctuations discussed in Chapter 8. According to the uncertainty principle, because the energy of the vacuum is fluctuating and never exactly zero, it can spontaneously create particles so long as they do not live longer than the uncertainty principle permits; this is why they are termed virtual. The expectation was that surrounding a point electron with a cloud of virtual particles would soften the singularities.

Vain hope. Matters got worse and infinities appeared everywhere. Mathematical methods known as *renormalization* were invented to cure the theory of infinities and give finite answers—which miraculously agree with experiment to such a precision that quantum electrodynamics is often called the most precisely tested theory ever created.

Originally, no one understood why renormalization worked.

Even one of its inventors, Richard Feynman, called it "hocus-pocus." Nowadays, the process is on firmer mathematical footing, but in any case renormalization is still considered essential for a viable field theory; if a theory cannot be renormalized to give sensible answers, it is discarded.

Unfortunately, not only do the infinities persist in standard attempts to quantize gravity, but the renormalization process fails and the theory cannot give sensible results.

That grave difficulty has resulted in a profusion of approaches toward creating a full theory of quantum gravity. The simplest avenue is to assume that gravity can be described classically, by general relativity, while treating any other fields in the problem, such as light, by the methods of quantum field theory. Physicists refer to such an approach as "semi-classical," which is a polite way of calling it a bastardized tactic. Nevertheless, it can be expected to bear fruit when the gravitational fields in the problem are not *too* strong—say, around large enough black holes. (The larger the black hole, the weaker its field.) For sure, the semi-classical approach resulted in quantum gravity's most famous triumph: Stephen Hawking followed this route to his celebrated 1974 discovery that black holes are not completely black but radiate energy, exactly the heat of black bodies.

Because it is so weak, black hole radiation has not been directly observed. That the temperature of a black hole of one solar mass would be about a ten-millionth of a degree, and the temperature of larger black holes even less, gives an idea of its feebleness. The fact, however, that Hawking's calculation showed that the radiation should be precisely that of a black body led most physicists to immediately accept the amazing result.

If black holes radiate energy, they must be losing mass. As they lose mass, their temperature increases; they emit energy more rapidly, losing mass more rapidly. This runaway effect led Hawking to predict that black holes would eventually end their lives in spectacular explosions. But his method actually assumes that the gravitational field, and thus the mass of the black hole, do not decrease. Such predictions, therefore, must be considered somewhat speculative. Indeed, the evaporation process should exert a feedback on the black hole such as to slow further evaporation; at least one of Hawking's colleagues claims, in fact, to have demonstrated that the feedback halts the evaporation long before any explosion takes place.

That result may turn out to be incorrect, but the example illustrates how difficult the issues are and how far we are from a full theory of quantum gravity. It is clear that Hawking's approach cannot be applicable at the Planck time.

*

What might be? Applicable at the Planck time, that is?

The most famous attack on the problem has been *string theory,* which lies beyond the scope of this little book. String theory attempts to be a unified field theory, or what is popularly called a *theory of everything*—a theory that not only unites the electromagnetic and nuclear forces (as do GUTS)—but includes gravity as well. String theory is a quantum field theory, but one in which the fundamental building blocks are not point particles; instead, they are tiny strings, whose length is approximately the Planck length. Once again, smearing points into finite strings might expunge infinities. The strings can have either open ends flapping about or be closed into loops. Ordinary particles are viewed as overtones of string vibrations, in the same way a violin string (or organ pipe) produces overtones.

A major difference between the strings of string theory and ordinary strings is that ordinary strings live in our universe of four spacetime dimensions (one time and three space), while, in one version of string theory, strings live in universes with ten spacetime dimensions (one time and nine space). The extra spatial dimensions are assumed to curl up on themselves, as around a cylinder, on lengths comparable to the Planck length. This is small enough so that we don't notice them.

String theory has had a number of mathematical successes. The most celebrated is that theorists have used it to derive the famous entropy of black holes, proposed by Jacob Bekenstein and made more precise by Hawking. (I won't talk about black hole entropy, but the result is famous and intimately related to the idea that black holes have a temperature.) String theory also predicts the particle that exchanges the gravitational force—the *graviton,* about which I'll say a little more shortly.

The appearance of the Planck length in string theory immediately tells us that it should indeed be a theory describing the extremely early universe. That is actually a severe difficulty; so far, string theory has made very little contact with other branches of physics. In particular, no earthbound experiment has been able to lend it any confirmation. What's more, the ten-dimension version is based on the concept from particle physics known as *supersymmetry,* which unites matter particles (like protons) with force particles (like photons) into a larger group. Not only is there no experimental evidence for supersymmetry, but results from the Large Hadron Collider seem to have all but ruled out the simplest versions.

Furthermore, the original attraction of superstring theory was that only one version of the theory appeared to be mathematically consistent. Nowadays, however, it is conceded that there may be 10^{500} different versions, a rather large proliferation of possibilities known as the *string-theory landscape*. The landscape should remind you of the multiverse from Chapter 12. One might reasonably argue that any theory that produces 10^{500} universes has not predicted anything. This is a serious issue.

*

Another attack on quantum gravity, not quite so well known as string theory, is *loop quantum gravity*. It does not intend to be a theory of everything but confines itself to quantizing gravity. It bears some resemblance to string theory in that its basic entities are loops, about the Planck length in size—but loop-gravity loops are fourdimensional. Indeed, they may be viewed not as existing in spacetime but as providing the basic building blocks of spacetime. Loop gravity calculations have also reproduced the Beckenstein-Hawking entropy of black holes.

In loop gravity, it simply does not make sense to talk about lengths smaller than the Planck length and times shorter than the Planck time; space and time themselves are quantized. It may help to visualize spacetime as a flexible lattice, whose bendable struts are of the Planck length and time. More closely, it probably resembles what, since long before the advent of loop gravity, has been popularly called *quantum foam*.

I have not emphasized the third important respect in which quantum mechanics differs from Newtonian physics, an aspect that goes hand in hand with the uncertainty principle. Quantum mechanics is a *probabilistic* theory. Unlike Newtonian mechanics, which tells us exactly where a particle will be in the future if we know its present position and velocity, quantum mechanics tells us only the probability that it will be in a certain place at a certain time.

It may be, then, that nothing so definable as "one centimeter" or "one second" exists in the Planck era. Quantum foam will require some probabilistic description that only "crystalizes" into our universe once the Planck era ends.

How would a quantum theory of gravity avert the singularity? Quantum fluctuations produce a pressure that manifests itself much like the repulsive force of the cosmological constant. If large enough it can bounce the universe during the Planck epoch. The exact results depend on the particular model being considered, which are too many to count. Loop quantum gravity claims to be able to do this, but no theory of quantum gravity has solved the cosmological constant problem—why today's cosmological constant is as small as it is.

One thing is nearly certain: to resemble our conventional field theories, in which forces are transmitted by particles, any theory of quantum gravity should predict the existence of a graviton, which would transmit the gravitational force. String theory does this. Although gravitational waves have been detected, however, individual gravitons have not and very likely never will be. If neutrinos interact with ordinary matter so rarely that one can pass through light-years of lead before hitting anything, then a graviton would interact with matter about twenty orders of magnitude *less* frequently, making direct detection of gravitons almost inconceivable.

This raises questions about how one could experimentally test a quantum theory of gravity. Some physicists feel it is not necessary that every facet of a theory be amenable to experiment. One might regard virtual particles as a mental or mathematical construct that helps us visualize how a field theory operates, although they are not directly detectable. What is important is that they predict phenomena that *are* directly detectable and confirm the theories.

On the other hand, if a theory predicts nothing that is directly detectable, then it has only mathematical consistency as an argument in its favor. As theories and models of the very early universe become ever farther removed from the realm of experience, some physicists argue that the traditional criterion for acceptance of a theory—that it be falsifiable, or capable of being proven wrong—is no longer tenable. Rather, we should be willing accept a theory on the basis of "meta-criteria," such as the probability that it is correct (if such a probability means anything) or even its artistic merits. To be sure, mathematical beauty has long been a driving force behind theory creation and acceptance, but proposals based on this elusive quality have turned out to be wrong as often as right.

So dramatically has the style and sociology of theoretical physics shifted in recent decades that the question inevitably arises: Have cosmologists taken to counting angels on pins? One also inevitably recollects the Yiddish proverb, "Man thinks and God laughs."

Have we entered an era of post-empirical science? Is post-empirical science an oxymoron?

MULTIVERSES AND METAPHYSICS

YOU HAVE BEEN PATIENTLY holding in reserve your question about the multiverse. I have been patiently waiting. After all, no cosmology lecture would be complete without its appearance. As for an answer, there is none better than the one James Peebles, America's grand old man of cosmology, gave after a 2020 Harvard talk. Did he believe in the multiverse?

No.

End of book.

In this case, it will be. As a rule, the press and the public are fascinated by the most extreme speculations and, as a rule, on a day-to-day basis, working cosmologists are not overly concerned with them. Nevertheless, the multiverse has been in the spotlight for well over a decade, and the excitement of pondering such matters is one reason that young people become cosmologists. As mentioned in Chapter 12 and Chapter 14, the inflationary model and string theory evidently require a multiverse.

But what is such a hyper-hydra-headed universe, exactly? "Exactly" may have no place in the question, or the answer. To an extent it is a matter of semantics. If by definition "universe" means "everything," then no multiverse exists. What is typically meant by "multiverse" in modern cosmology is an ensemble of "subuniverses" with wildly differing properties. Some may be flat; most will be curved. In some, the fundamental constants of nature will be at or near the values we measure them to be. In others they will be different by orders of magnitude. In some, galaxies will exist, in others not. We live in one of them.[①]

The multiverse is the epitome of "postempirical" science—there seems to be no way to test the multiverse concept by the traditional scientific methods sketched in the Introduction. A few proposals have been made, but none have been taken seriously enough to be actively pursued. Cosmologists search for dark matter because there is indirect observational evidence for it, but they are not searching for the multiverse, because there

---

① There is another sort of multiverse, associated with quantum mechanics. Quantum mechanics does not predict the outcome of a measurement, only the probability of a given outcome. Some physicists believe that at every measurement the universe splits, so that all outcomes occur, but in different universes. This is known as "the many worlds interpretation of quantum mechanics."

is no evidence for it. In his answer to the after-lecture question, Peebles reflected this position.

To be indulgent, we might ask why we are living in the particular universe we are. More specifically: Why do we observe our universe to be approximately ten billion years old?

This is the basic *anthropic* question. Robert Dicke's answer is famous: "The universe must have aged sufficiently for there to exist elements other than hydrogen, since it is well known that carbon is required to make physicists." In other words, if the universe weren't at least several billion years old, we wouldn't be here to observe it. More broadly the *anthropic principle* holds that the universe as we observe it must be such as to allow life. A universe that did not produce life would not produce observers. According to the anthropic principle, the existence of life selects our particular cosmos from the multiverse.

When anthropic arguments became popular in the 1970s, reactions ranged from skepticism to scorn. Many physicists dismissed it as tautological; *obviously* our universe is such as to be compatible with life. An analogy offered by Dicke and Peebles, though, may make it seem less trivial. Loaded and unloaded pistols are randomly distributed to a crowd of cosmologists and they engage in a mass game of Russian roulette. Afterward, a brilliant statistician appears and discovers by exhaustive analysis that there is a high probability that any surviving cosmologist holds an unloaded pistol.

You might derisively exclaim, "Obviously!" That outcry, however, is an admission that the situation is subject to meaningful after-the-fact analysis. The main objection to the anthropic principle has always been that it cannot predict anything, and therefore fails at the fundamental requirement of a physical theory. The mass Russian roulette game makes that a little less clear; the outcome *could* have been predicted. When playing roulette with universes, admittedly, there is no way to know ahead of the game whether a given universe is loaded.

Nevertheless, a famous story in anthropic lore is that astronomer Fred Hoyle used anthropic reasoning in 1953 to predict that a certain nuclear reaction in the sun *must* exist for sufficient carbon to be produced to sustain life. Nowhere in his papers of the time, however, does he mention anthropic considerations, and the story appears to be a retrospective invention.

The situation differs with regard to American geologist Thomas Chamberlin. In the nineteenth century, a great debate raged between physicists and naturalists over the age of the earth. Darwin required untold eons to evolve the species, but physicists led by Lord Kelvin did not believe that the sun could have lasted long enough to do so, emitting energy by any known mechanism. In 1899, Chamberlin argued that Kelvin's arguments proved only that the sun was burning by some unknown source of energy locked in atoms. The Darwinians and Chamberlin turned out to be correct and the physicists wrong. Chamberlin's reasoning might have led to the discovery of nuclear reactions in the sun.

In recent decades, anthropic arguments have been enlisted to explain numerous features of our universe, although only in hindsight. Most relevant to our purposes are the arguments to constrain the sizes of the fingerprints of God and the cosmological constant. We have seen that the size of the fluctuations in the microwave background are roughly one part in 10^5. If they were much larger, the matter in the universe would have collapsed into black holes. If they were much smaller, the matter would not have coalesced into galaxies and stars. In neither case would observers have arisen in such a universe.

By the same token, because it accelerates the expansion of the universe, the cosmological constant impedes matter from coalescing into galaxies. If the constant were larger than the matter content of the universe during the epoch of galaxy formation, when the observable universe was roughly a fifth its current size, no galaxies could form. The density of matter back then was about 125 times larger than at present, and so presumably the cosmological constant could not have been more than one or two orders of magnitude larger than it is today.

A major objection to anthropic arguments has always been that they rarely yield an answer with a give-or-take range less than an order of magnitude. True. On the other hand, limiting the cosmological constant to a factor of ten or so above its present value is a significant improvement over the 120 orders of magnitude mentioned in Chapter 8 based on quantum mechanical calculations.

Many physicists, even those who propose them, regard anthropic arguments as acts of desperation. They may be inevitable in an age when our quantitative theories have become so speculative; it is wrong to think that a theory filled with complicated equations necessarily means anything. One should also bear in mind that the anthropic principle is a *principle,* not a law of nature. Many principles have been enlisted throughout the history of physics to guide our thinking toward successful theories; some have proven more useful than others. The cosmological principle proved to be very successful, even if it was obviously not entirely true. But how does one test the principle of beauty? The idea of beauty in physics is often encapsulated in the idea of mathematical symmetry—that systems have regular patterns—and while the implementation of symmetry concepts has proven highly successful in particle physics, it may have outlived its usefulness. As mentioned in Chapter 14, the Large Hadron Collider has found no evidence for supersymmetry.

A famous *principle of least action* is universally accepted among physicists. The principle of least action springs from the simple idea that the shortest distance between any two points is a straight line and that light, for example, tends to travel along those straight lines. The principle says that one can obtain the equations for a given theory by minimizing a quantity known as the *action,* which is related to a system's energy. Historically, the action principle revolutionized physics and has become the route by which *all* modern theories are created. Rather than infer the correct equations from experience, one postulates an action and minimizes it to generate the theory's equations. Einstein did not consider his general relativity theory complete until he could derive the field equations from an action.[①] Theories of quantum

---

① Mathematician David Hilbert beat Einstein in this race by five days.

gravity also begin by postulating an action. It is known, however, that sometimes the principle of least action gives the wrong answer. If we are merely postulating an action for a completely new theory, how do we know we have produced the correct equations? Especially when we cannot experimentally test the results?

On a spectrum ranging from the principle of beauty to the principle of least action, the anthropic principle perhaps lies closer to beauty. Moreover, I have been discussing what is known as the *weak* version of the anthropic principle, which if not tautological, does not seem unreasonable. As in Dicke's original question, it merely asks why some aspect of the universe—its age—is observed to be what it is. It assumes that the known laws of nature are what they are. Stronger versions of the anthropic principle declare that the laws of nature *must* be as they are. In particular, that the fundamental constants of nature, such as G and h, must be what we measure them to be; otherwise the universe as we know it could not exist. For example, if the constants were much different than they actually are, stars would not form, and so presumably life would not exist, either.

Physicists have a harder time accepting the strong anthropic principle because it holds echoes of the argument from design—the great clockwork of the universe must imply the existence of a watchmaker. To be sure, the strongest version of the anthropic principle, the *participatory* anthropic principle, requires the universe to eventually produce life. Physicists generally reject such ideas because they smack of teleology—the belief that things happen because of the final purpose they serve. Science has moved in the opposite direction from teleological arguments since Aristotle.

✷

Without enlisting the anthropic principle, it is presently unclear how to select viable universes from the multiverse or from the string theory landscape. The current state of affairs is undoubtedly due to the lack of experiments or observations constraining the imagination of theorists. Even if we are lucky enough to become an advanced civilization, it will remain a stretch to create universes in the laboratory to test the multiverse and the anthropic principle.

It is likely that we will never completely understand what took place at the Planck time, or before the big bang, unless the newer bounce cosmologies allow us to peer into that epoch. If our theories do not in the end provide a smooth transition to what is generally observable, we may indeed be forced to rely on mathematical consistency, and vague notions of probability and beauty to constrain them.

By the same token, it is unlikely that physicists will ever create a theory of everything. The phrase should not be taken too seriously. Even those attempting to create one would not claim it could explain why people fall in love. However, even in its more limited goal, to unite the four forces of nature, it is hardly clear how useful it would be. Along the path to a theory of everything many insights have been achieved, but a large number of scientists regard the endeavor as misguided in principle.

The most successful theories are those with limited domains of applicability. No knowledge of what took place in the earliest instants of the universe is needed to calculate the orbits of the planets. Perhaps the greatest achievement of science is that it is possible to say something without saying everything. And there is no question that a theory of everything would remain incomplete. A ten-dimensional string theory, even if accepted beyond a shadow of a doubt, would leave unanswered the question of why there are ten dimensions. No theory specifies everything about itself. Whether it is the very constants of nature, or assumptions of how the universe began, something always remains to be put in by hand. Most cosmologists would concede that they do not study cosmology to solve the ultimate mysteries of nature but to get close to them. Rest easy, then, and do not worry: future generations of cosmologists will continue to wonder . . .

Why is there something rather than nothing?

ACKNOWLEDGMENTS

My deep thanks to Stephen Boughn and Patti Wieser for critical readings of the book. Thanks also to the anonymous reviewers for their helpful suggestions. Of course, any remaining mistakes or oversights are my own responsibility.